Firefighter Fatalities in the United States in 2000

I0470174

Prepared for

National Fire Data Center
United States Fire Administration
Federal Emergency Management Agency
Contract Number EME-1998-CO-0202-T009

Prepared by

IOCAD Emergency Services Group

Firefighter Fatalities in the United States in 2000

U.S. Fire Administration - Mission Statement

As an entity of the Federal Emergency Management Agency, the mission of the United States Fire Administration is to reduce life and economic losses due to fire and related emergencies, through leadership, advocacy, coordination and support. We serve the Nation independently, in coordination with other Federal agencies, and in partnership with fire protection and emergency service communities. With a commitment to excellence, we provide public education, training, technology and data initiatives.

Firefighter Fatalities in the United States in 2000

Firefighter Fatalities in the United States in 2000

ACKNOWLEDGMENTS

This study of firefighter fatalities would not have been possible without the cooperation and assistance of many members of the fire service across the United States. Members of individual fire departments, chief fire officers, the National Interagency Fire Center, United States Forest Service personnel, the United States military, the Department of Justice, NFPA International, and many others contributed important information for this report.

IOCAD Emergency Services Group of Emmitsburg, Maryland (a division of IOCAD Technical Services, Inc.) conducted this analysis for the United States Fire Administration (USFA) under contract EME-1998-CO-0202-T009.

The ultimate objective of this effort is to reduce the number of firefighter deaths through an increased awareness and understanding of their causes and how they can be prevented. Fire fighting, rescue, and other types of emergency operations are essential activities in an inherently dangerous profession, and unfortunate tragedies do occur. This is the risk all firefighters accept every time they respond to an emergency incident. However, the risk can be reduced greatly through efforts to increase firefighter health and safety.

Photographic Acknowledgments

The United States Fire Administration wishes to thank the following sources for contributing the photographs used in this report:

- The Deseret News and photographer Laura Seitz for the use of the photograph that appears on the cover. The photograph depicts the funeral procession of Firefighter Kendall O. Bryant of the Layton City, Utah, Fire Department who died on March 31, 2000.

- The State Journal-Register of Springfield, Illinois, and photographer Shawn Poynter for the use of the photograph that appears on the cover which depicts the funeral procession of Captain Steven Wilmot of the Springfield Fire Department who died on August 9, 2000.

- The News-Press of Fort Myers, Florida, and photographer Darron R. Silva for the use of the photograph that appears on page 10. The photograph shows the scene of a helicopter crash that claimed the life of Firefighter/Rotor Craft Pilot George A. "Bo" Burton of the Florida Division of Forestry who died on June 4, 2000.

- The Fayetteville Observer of Fayetteville, North Carolina, and photographer Marcus Castro for the use of the photograph that appears on page 15. The photograph shows the scene of a collision between a ladder truck and a train that claimed the life of Firefighter/Engineer David Clements Sharp, II, of the Fayetteville Fire/Emergency Management Department who died on March 17, 2000.

This report is dedicated to the families of those firefighters who made the ultimate sacrifice in 2000. May the lessons learned from their passing not go unheeded.

Firefighter Fatalities in the United States in 2000

BACKGROUND

For more than 20 years, the United States Fire Administration (USFA) has tracked the number of firefighter fatalities and conducted an annual analysis. Through the collection of information on the causes of firefighter deaths, the USFA is able to focus on specific problems and direct efforts toward finding solutions to reduce the number of firefighter fatalities in the future. This information also is used to measure the effectiveness of current programs directed toward firefighter health and safety.

One of the USFA's main program goals is a 25 percent reduction in firefighter fatalities in 5 years and a 50 percent reduction within 10 years. The emphasis placed on these goals by the USFA is underscored by the fact that these goals represent one of the four major objectives that guide the actions of the USFA.

In addition to the analysis, the USFA provides a list of firefighter fatalities to the National Fallen Firefighters Foundation. If Memorial criteria are met, the fallen firefighter's next of kin, as well as members of the individual fire department, are invited to the annual Fallen Firefighters Memorial Service. The service is held at the National Emergency Training Center in Emmitsburg, Maryland, annually during Fire Prevention Week. Additional information regarding the Memorial Service can be found on the internet at http://www.firehero.org/ or by calling the National Fallen Firefighters Foundation at (301) 447-1365. An updated list of firefighter fatalities from 1981 through the present, including a searchable database, may be found at
http://www.usfa.fema.gov/ffmem/ffmem_search.cfm

INTRODUCTION

This report continues a series of annual studies by the USFA of on-duty firefighter fatalities in the United States.

The specific objective of this study was to identify all on-duty firefighter fatalities that occurred in the United States in 2000, and to analyze the circumstances surrounding each occurrence. The study is intended to help identify approaches that could reduce the number of firefighter deaths in future years.

In addition to the 2000 overall findings, this study includes assessments of trends over the past 5 years and special analyses on heart attacks and ways to immediately prevent future firefighter deaths.

Who is a Firefighter?

For the purpose of this study, the term firefighter covers all members of organized fire departments in all States, the District of Columbia, and the Territories of Puerto Rico, Virgin Islands, American Samoa, Commonwealth of the Northern Mariana Islands, and Guam, including career and volunteer firefighters; full-time public safety officers acting as firefighters; State, Territory, and Federal government fire service personnel, including wildland firefighters; and privately employed firefighters, including employees of contract fire departments and trained members of industrial fire brigades, whether full- or part-time. It also includes contract personnel working as firefighters or assigned to work in direct support of fire service organizations.

Firefighter Fatalities in the United States in 2000

Under this definition, the study includes not only local and municipal firefighters, but also seasonal and full-time employees of the United States Forest Service, the Bureau of Land Management, the Bureau of Indian Affairs, the Bureau of Fish and Wildlife, the National Park Service, and State wildland agencies. The definition also includes prison inmates serving on fire fighting crews; firefighters employed by other governmental agencies such as the United States Department of Energy; military personnel performing assigned fire suppression activities; and civilian firefighters working at military installations.

What Constitutes an On-Duty Fatality?

On-duty fatalities include any injury or illness sustained while on-duty that proves fatal. The term on-duty refers to being involved in operations at the scene of an emergency, whether it is a fire or non-fire incident; responding to or returning from an incident; performing other officially assigned duties such as training, maintenance, public education, inspection, investigations, court testimony, and fundraising; and being on-call, under orders, or on standby duty, except at the individual's home or place of business. An individual who experiences a heart attack or other fatal injury at home as he or she prepares to respond to an emergency is considered on-duty when the response begins.

A fatality may be caused directly by an accidental or intentional injury in either emergency or non-emergency circumstance, or it may be attributed to an occupationally-related fatal illness. A common example of a fatal illness incurred on-duty is a heart attack. Fatalities attributed to occupational illnesses also would include a communicable disease contracted while on-duty that proved fatal when the disease could be attributed to a documented occupational exposure.

Injuries and illnesses are included when death is delayed considerably after the original incident. When the incident and the death occur in different years, the analysis counts the fatality as having occurred in the year that the incident took place.

There is no established mechanism for identifying fatalities that result from illnesses that develop over long periods of time, such as cancer, that may be related to occupational exposure to hazardous materials or products of combustion. It has proved to be very difficult over several years to provide a full evaluation of an occupational illness as a causal factor in firefighter deaths. This situation is attributable to the limitations in the ability to track the exposure of firefighters to toxic hazards, the often delayed long-term effects of such exposures, and the exposures firefighters may receive while off-duty.

Firefighter Fatalities in the United States in 2000

Sources of Initial Notification

As an integral part of its ongoing program to collect and analyze fire data, USFA solicits information on firefighter fatalities directly from the fire service and from a wide range of other sources. These sources include the Public Safety Officers' Benefit Program (PSOB) administered by the Department of Justice, the National Institute for Occupational Safety and Health (NIOSH), the Occupational Safety and Health Administration (OSHA), the United States military, the National Interagency Fire Center, and other Federal agencies.

The USFA receives notification of some deaths directly from fire departments, as well as from such fire service organizations as the International Association of Fire Chiefs (IAFC), the International Association of Fire Fighters (IAFF), NFPA International, the National Volunteer Fire Council (NVFC), State fire marshals, State training organizations, other State and local organizations, fire service internet sites, news services, and fire service publications. The USFA also keeps track of fatal fire incidents as part of its Major Fire Investigations Program and performs an ongoing analysis of data from the National Fire Incident Reporting System (NFIRS).

Procedure for Including a Fatality in the Study

In most cases, after notification of a fatal incident, initial telephone contact is made with local authorities by the USFA's contractor to verify the incident, its location and jurisdiction, and the fire department or agency involved. Further information about the deceased firefighter and the incident may be obtained from the chief of the fire department or his or her designee over the phone or by other data collection forms.

Information that is requested routinely includes NFIRS-1 (incident) and NFIRS-3 (fire service casualty) reports, the fire department's own incident reports and internal investigation reports, copies of death certificates or autopsy results, special investigative reports, police reports, photographs and diagrams, and newspaper or media accounts of the incident. Information on the incident also may be gathered from NFPA International, the USFA, or NIOSH reports on an incident.

After obtaining this information, a determination is made as to whether the death qualifies as an on-duty firefighter fatality according to the previously described criteria. The same criteria were used for this study as in previous annual studies. Additional information may be requested, either by followup with the fire department directly, from State vital records offices, or other agencies. The determination as to whether a fatality qualifies as an on-duty death for inclusion in this statistical analysis is made by the USFA. The final determination as to whether a fatality qualifies as a line-of-duty death for inclusion in the Fallen Firefighters Memorial Service is made by the National Fallen Firefighters Foundation.

3

Firefighter Fatalities in the United States in 2000

2000 FINDINGS

One hundred and two firefighters died while on-duty in 2000. This represents a decrease of 10 deaths from 1999. The total of 102 firefighter fatalities will hopefully reflect a movement back to the overall downward trend of firefighter deaths that had been experienced in the 1990s with the exception of 1999.

The total of 102 fatalities is the fourth-time in the last 10 years when the total number of firefighter fatalities has exceeded 100. The lowest years on record were 1992 with 75 fatalities and 1993 with 77 fatalities.

Figure 1 - On-Duty Firefighter Fatalities (1977-2000)

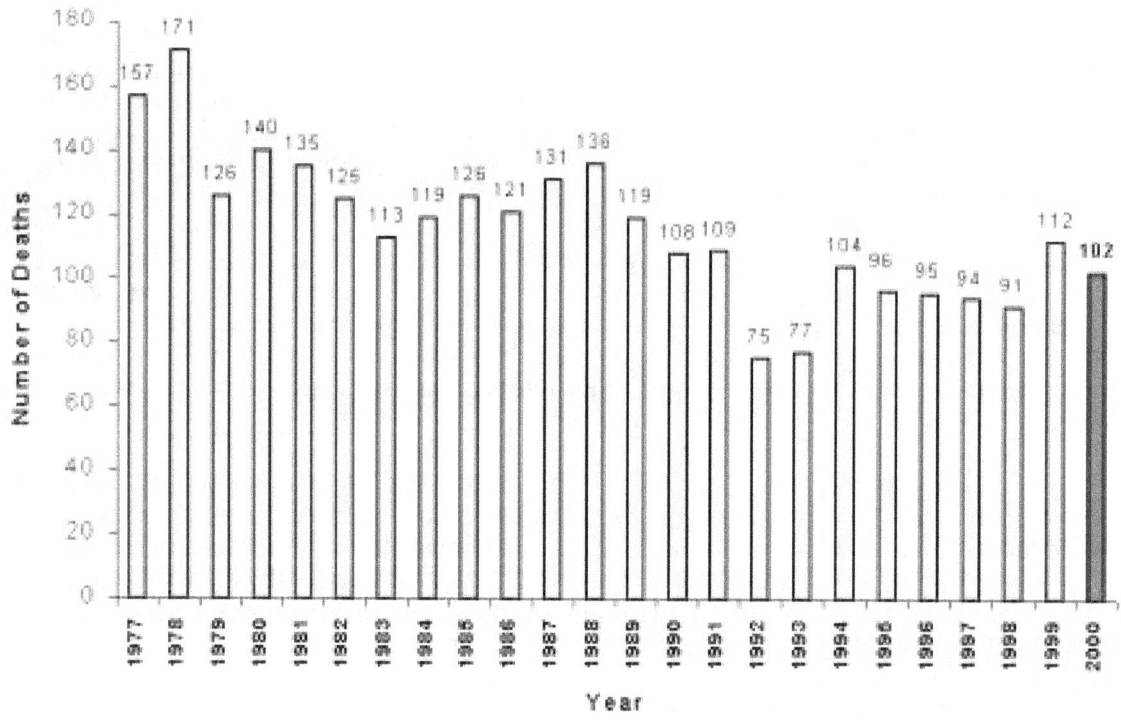

While lower than the 1999 total, this year's total does not return completely to the gradual slowing in the firefighter death rate that had been experienced in the latter part of the 1990s. Despite the losses in 1999 and 2000, the 10-year trend in firefighter fatalities is down 23 percent. However, the rate of firefighter fatalities in the last 5 years has risen 10 percent, attributable partly to the uncharacteristically low number of deaths that occurred in 1992 and 1993 (Figure 1) and the higher number of firefighter fatalities in 1999 and 2000.

Firefighter Fatalities in the United States in 2000

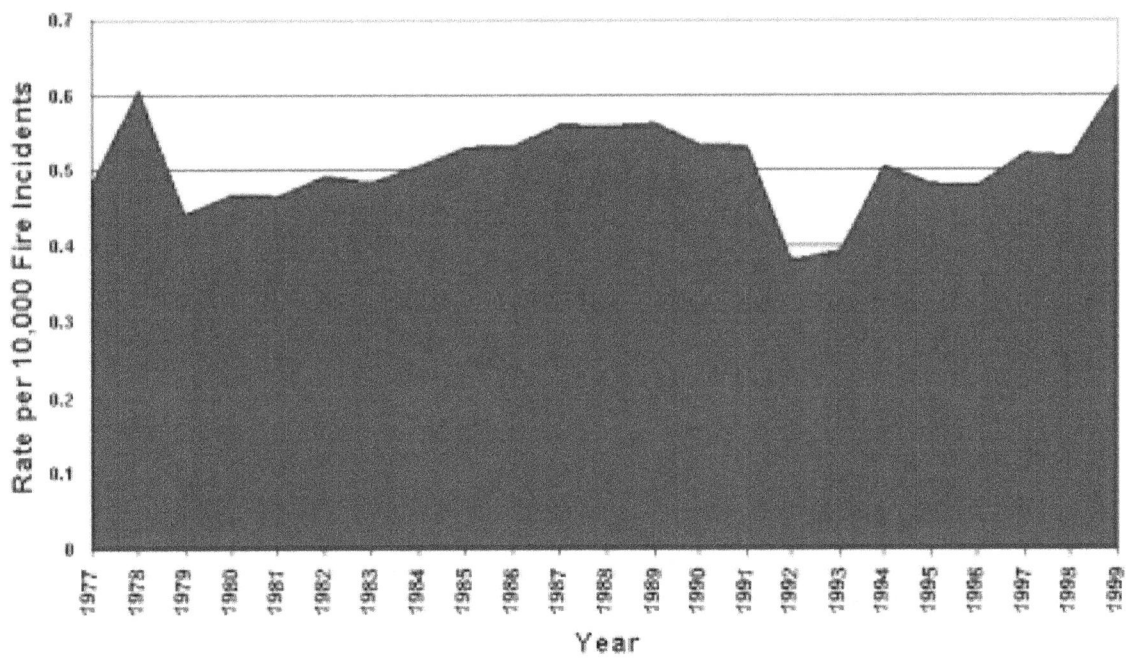

Firefighter Fatalities per 10,000 Fire Incidents
1977 - 1999

While the total number of firefighter fatalities have been trending downward over the past 20 years, the number of firefighter deaths per fire incident may have risen. The chart above compares the total number of firefighter fatalities each year and the total number of fire incidents reported by NFPA International through 1999. While firefighters die in many non-fire situations and deaths are compared to fire incidents only in the figure above, the information presented above suggests that fire fighting is getting more hazardous. A retrospective study of firefighter fatalities that covers a period of 10 years or more is being conducted at this time and will shed more light on this possibility. The retrospective study of firefighter fatalities is scheduled to be released in late 2001.

The 2000 firefighter fatalities included 64 volunteer firefighters and 38 career firefighters (Figure 2). Among the volunteer firefighter fatalities, 59 were from local or municipal volunteer fire departments, and 5 were seasonal or contract members of wildland fire agencies. Of the career firefighters who died, 30 were members of local or municipal fire departments, 7 were career members of wildland fire fighting agencies, and 1 was a career military firefighter. Ninety-nine of the fatalities were men and 3 were women.

Firefighter Fatalities in the United States in 2000

Figure 2
Career vs. Volunteer Deaths

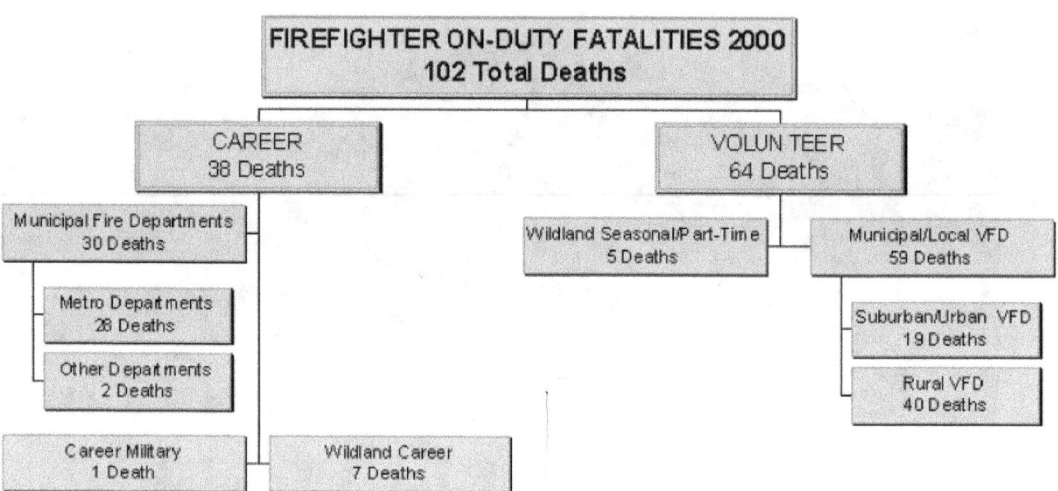

The average years of experience for a firefighter that died in 2000 was 15 1/2.

Table 1.
Multiple Firefighter
Fatality Incidents

Year	Number of Incidents	Total Number Deaths
2000	5	10
1999	6	22
1998	10	22
1997	8	17
1996	3	8

Multiple Firefighter Fatality Incidents

The 102 deaths resulted from 97 incidents. There were 5 multiple fatality incidents resulting in the deaths of 10 firefighters.

In 2000, 2 Texas firefighters died when they were trapped by a roof collapse in a restaurant fire; 2 Tennessee firefighters were killed when they were ambushed and killed by a gunman as they arrived on the scene of a single-family residential fire; 2 firefighters were killed in an aircraft crash in New Mexico; 2 Utah firefighters were killed when they were struck by lightning as they took shelter under some trees from a sudden storm; and 2 North Carolina firefighters were killed in the crash of their helicopter.

Firefighter Fatalities in the United States in 2000

Wildland Fire Fighting Deaths

The number of deaths associated with brush, grass, or wildland fire fighting in 2000 was 18. In 2000, there were 6 firefighter deaths associated with aircraft fire fighting duties (2 other aircraft deaths occurred during non-fire fighting duties and are not included in the 18). This total includes fixed-wing aircraft and helicopters. In 1999, there were no firefighter fatalities resulting from aircraft crashes.

Two Utah firefighters were killed when they were struck by lightning as they took shelter under some trees from a storm; 3 firefighters were killed in separate incidents when their positions were overrun by fire progress; 2 firefighters were killed in training associated with wildland fire fighting, 1 died of a heart attack after cutting a fire break, and a smokejumper died when his parachute failed to open during a training jump. One firefighter was mortally burned when a pile of fire debris that was thought to be extinguished "blew up"; 1 firefighter was killed after extinguishing a grass fire that resulted from the use of a cutting torch on an obsolete fuel tank that was being cut up for salvage, further cutting on the tank produced an explosion; 1 firefighter was killed while responding to a grass fire in a brush truck when a sedan crossed the centerline of the roadway and impacted the brush truck; 1 firefighter suffered a heart attack at a grass fire; and 1 firefighter was killed by burns suffered when his All Terrain Vehicle (ATV) rolled over during a controlled burn. Gasoline leaked from the vehicle's fuel tank onto the firefighter and the fuel was ignited by a drip torch being carried on the ATV.

Table 2.
Firefighter Deaths Associated with Wildland Fire Fighting

Year	Total Number Deaths
2000	18
1999	28
1998	13
1997	9
1996	5

Firefighter Fatalities in the United States in 2000

In 2000, ten firefighters were killed while working on the scene of arson caused fires.

Table 3.
Emergency Duty
Firefighter Deaths

Year	Percentage of all Deaths
2000	71%
1999	87%
1998	77%
1997	81%
1996	72%

Type of Duty

In 2000, 72 on-duty firefighter deaths were associated with emergency incidents, accounting for 71 percent of the 102 fatalities (Figure 3). This includes all firefighters who died while responding to an emergency, while at the emergency scene, or while returning from the emergency incident. Non-emergency activities accounted for 30 fatalities (29 percent). Non-emergency duties include training, administrative activities, or performing other functions that are not related to an emergency incident. A 5-year historical perspective concerning the percentage of firefighter deaths that occurred during emergency duty is presented in Table 3.

Figure 3 - Firefighter Deaths by Type of Duty (2000)

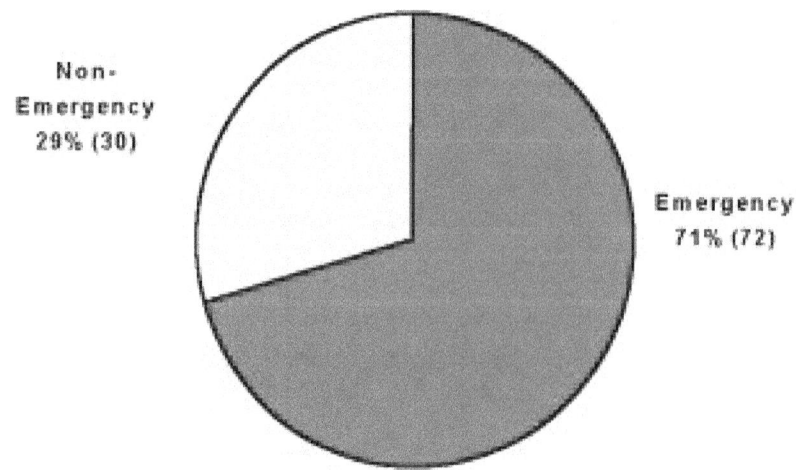

Non-Emergency 29% (30)

Emergency 71% (72)

Firefighter Fatalities in the United States in 2000

The number of deaths by type of duty being performed in 2000 is shown in Table 4 and presented graphically in Figure 4. As in previous years, the largest number of deaths occurred during fireground operations. There were 41 fireground deaths, which accounted for 40.2 percent of the fatalities, down from 54 percent in 1999, 46 percent in 1998, and 44 percent in 1997. Of the 41 fireground deaths, slightly more than 1 in 3 (14) resulted from heart attacks that occurred on the fire scene. Other fireground deaths included 8 from asphyxiation, 9 from internal trauma, 6 from burns, and 2 electrocutions (lightning). One firefighter died of pneumonia the night that he was discharged after a three-day hospital stay caused by a fall at a fire scene, and 1 firefighter experienced a CVA (stroke) at an automatic fire alarm incident.

Table 4.
2000 Firefighter Deaths by Type of Duty

Type of Duty	Number of	Percent of Deaths
Fireground Operations	41	40.2%
Responding/Returning from Alarm	19	18.6%
Other On-Duty	16	15.7%
Training	13	12.7%
Non-Fire Emergencies	11	10.8%
After an Incident	2	2%
Total	102	100%

Figure 4 - Fatalities by Type of Duty (2000)

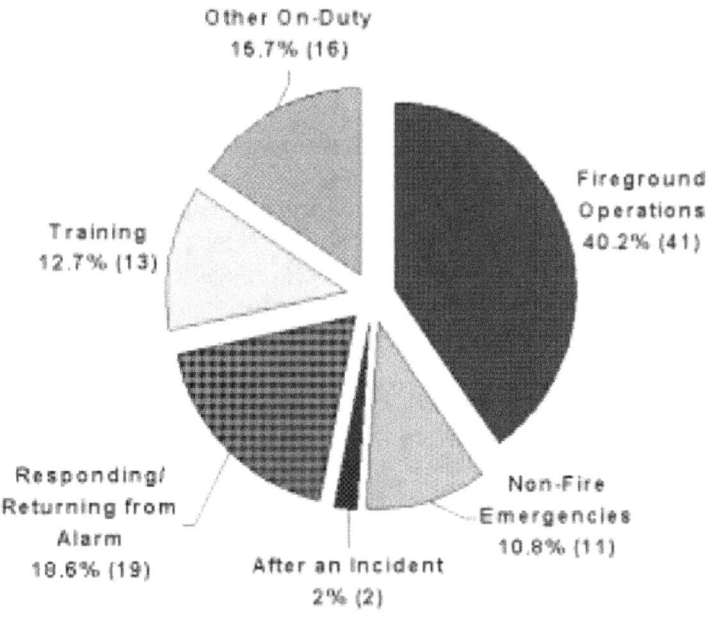

Nineteen firefighters died while responding to or returning from emergency incidents. This has been the second leading type of duty in which firefighter deaths have occurred each year since 1993. In 2000, 16 of the 19 firefighter deaths that occurred while responding to or returning from an incident involved volunteer firefighters.

Firefighter Fatalities in the United States in 2000

In 2000, twelve firefighters died while responding to incidents and seven died while returning from incidents.

Of the 19 firefighters who died while responding to or returning from alarms, 13 died of traumatic injuries and 6 died of heart attacks. The traumatic injuries consisted of 3 firefighters who died in collisions involving their personal vehicles while responding to emergencies; 1 firefighter that was killed when he was ejected from a ladder truck after a collision; 1 firefighter that was killed when he slipped and was crushed by a ladder truck that he was chasing as it began its response; 1 firefighter that was killed while returning to quarters when his ladder truck was struck by a train at a crossing; 6 firefighters who were killed in collisions involving fire apparatus; and 1 firefighter that was killed in the crash of a helicopter that was returning to base.

There were 6 firefighter deaths due to heart attacks while responding to or returning from alarms. Three firefighters died of heart attacks immediately after returning home from an emergency incident; 1 firefighter suffered a fatal heart attack as he drove a rescue truck to an incident; 1 firefighter died of a heart attack he suffered while traveling from a fire scene to the fire station to retrieve a piece of fire apparatus; and 1 firefighter experienced a heart attack in his personal vehicle only blocks from his fire station after returning from an incident.

There were 16 deaths that occurred during other on-duty activities. These deaths include 7 firefighters who died from heart attacks while on-duty or in the fire station and 1 who died from an internal hemorrhage while on-duty. Two firefighters were killed when their helicopter crashed en-route to a public education event, and 2 firefighters were killed in separate incidents after they fell while working on fire station garage door openers. One firefighter was killed when a fire hose struck him during testing, the force of the water propelled him into a piece of fire apparatus and he sustained a fatal head injury. A career military firefighter was killed when he was struck by another Aircraft Rescue Fire Fighting (ARFF) vehicle as he conducted a morning pump test, and 1 firefighter was killed when he was struck as he crossed the road to retrieve chains that were going to be used to tow a disabled fire truck. One firefighter was shot to death as he participated in a "fill the boot" fund drive.

In 2000, 13 firefighters died during training exercises. This number is significantly higher than the loss that was experienced in 1999. The firefighter deaths in 2000 included 7 heart attacks: Four firefighters suffered heart attacks as they

Firefighter Fatalities in the United States in 2000

participated in structural fire fighting training - 1 after wildland training which involved cutting fire breaks, 1 after completion of a physical fitness test, and 1 during a physical fitness workout. Two firefighters drowned during dive rescue training; 1 firefighter was trapped by fire progress in a training fire in an abandoned structure; 1 firefighter was killed in a tanker (tender) rollover during shuttle training; and 1 firefighter was killed during smokejumper training when his parachute failed to open during a jump. One firefighter was killed as he participated in SWAT-medic training when a lead plug in the barrel of a practice weapon dislodged when the weapon was fired. The plug struck the firefighter in the head and delivered fatal injuries.

Eleven deaths were related to activities at the scene of non-fire emergency incidents. Heart attacks were the leading killer with 6 occurring during non-fire emergencies in 2000. Two fire police officers died in separate incidents when they suffered heart attacks on the scene of motor vehicle collisions; 1 firefighter suffered a heart attack as he drove a fire department ambulance that was transporting a patient to the hospital; 1 firefighter suffered a heart attack in the patient compartment of an ambulance while transporting a patient; 1 firefighter suffered a heart attack at the scene of an extended vehicle extrication operation; and 1 firefighter suffered a heart attack as his company assisted the police with the ventilation of the site of a previous civilian death. One firefighter drowned when he was drawn into a rain swollen drainage culvert and pipe as he attempted to rescue a civilian from the water, 3 firefighters were killed in separate incidents when they were struck by vehicles at the scene of previous vehicle collisions, 1 firefighter suffered a CVA (stroke) while on storm callout during tornado activity.

Two firefighters died in 2000 after the conclusion of an incident. One firefighter died after he fell on the scene of a fire investigation and suffered internal injuries. The injuries were not diagnosed and caused the firefighter's death 22 days after the injury. One firefighter died of a heart attack after responding to a motor vehicle collision.

**Table 6.
Firefighter Deaths
During Training**

Year	Number of Firefighter Deaths
2000	13
1999	3
1998	12
1997	5
1996	6

Firefighter Fatalities in the United States in 2000

Career, Volunteer, and Wildland Deaths by Type of Duty

Figure 5 depicts career, volunteer, and wildland firefighter deaths by type of duty. Wildland career, wildland seasonal, and wildland contractor deaths were grouped together. This chart demonstrates the disproportionate number of fatalities experienced by volunteer firefighters responding to and returning from alarms as compared to career and wildland firefighters. In 2000, 25 percent of volunteer firefighter deaths were while responding to or returning from emergencies. This compares to 9.7 percent of the career deaths and 8 percent of the wildland deaths. This is a continuing trend.

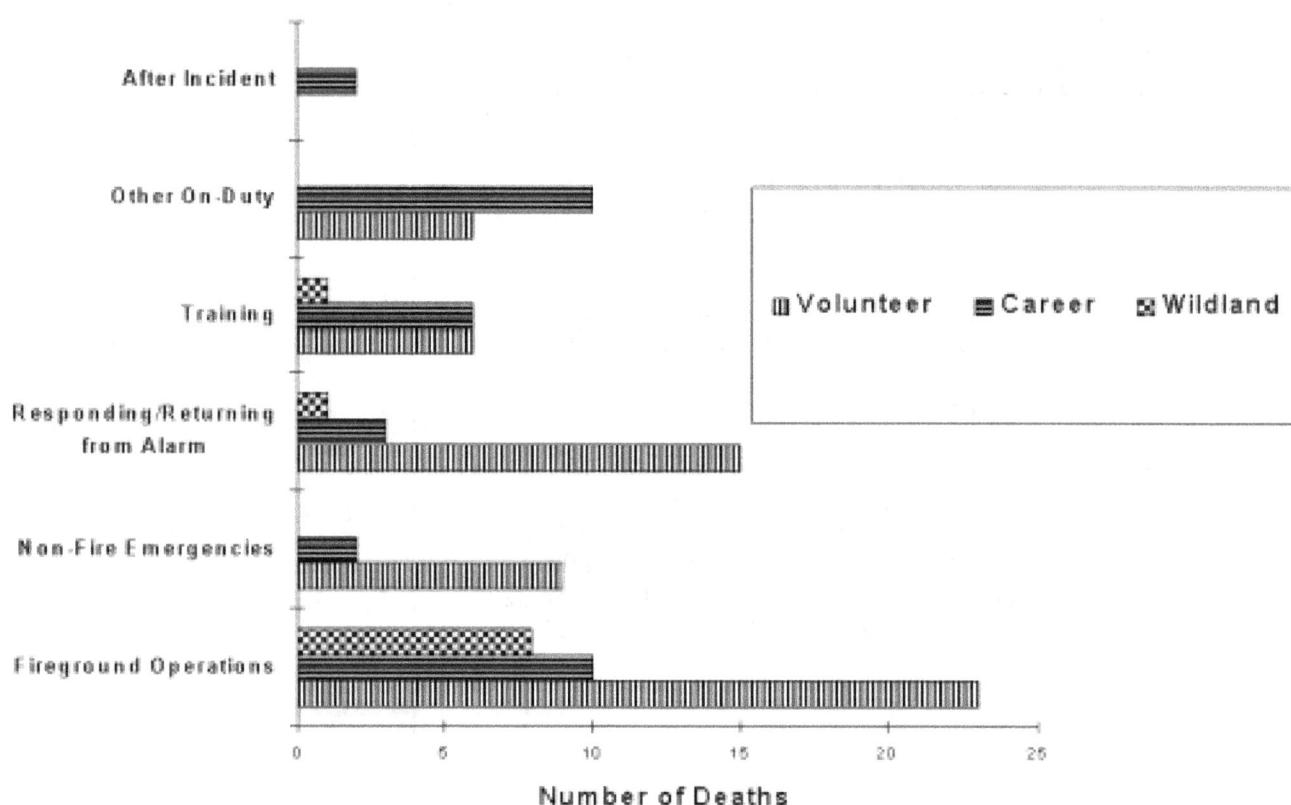

Figure 5 - Career, Volunteer, and Wildland Deaths by Type of Duty (2000)

Firefighter Fatalities in the United States in 2000

The large number of career firefighter deaths while on-duty but not involved in an incident or training activity may be attributed to the fact that career firefighters are on-duty for longer periods of time than volunteer firefighters. The on-duty periods for volunteer firefighters generally are related to an emergency incident or other official functions such as training. Some volunteer fire departments staff stations overnight (similar to a career department) but their numbers are small when compared to the total number of volunteer fire departments.

Type of Emergency Duty

Sixty-five firefighters died while engaged directly in emergency service delivery, including deaths that were the result of injuries sustained on the incident scene or en-route to the incident scene. This total includes firefighters that became ill on an incident scene and later died but does not include firefighters who became ill or died while returning from an incident (such as a vehicle collision while returning from an incident). Figure 6 shows the percentage of firefighters killed in fire fighting, emergency medical services, technical rescue-related incidents, and other emergency incidents. Fifty-two firefighters were killed in relation to fires, 11 in relation to EMS calls, 1 while engaged in a technical rescue, and 1 of a CVA (stroke) as he performed tornado storm watch duties.

Figure 6 - Type of Emergency Duty (2000)

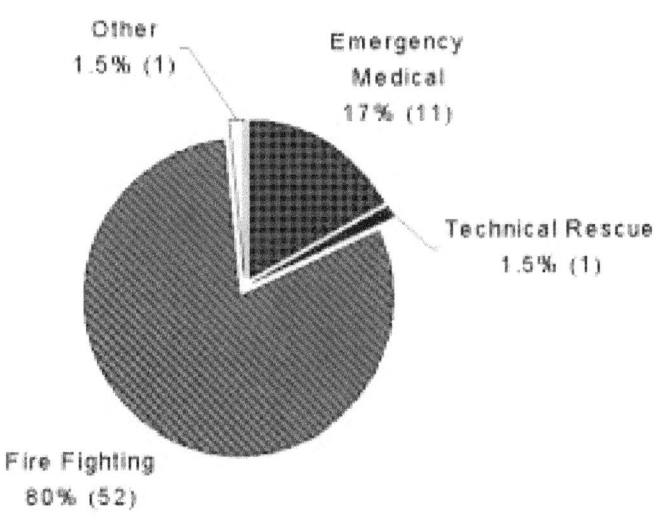

Note - 65 of 102 – On-Scene or Responding

Firefighter Fatalities in the United States in 2000

Table 7.
Cause of Fatal Injury - 2000

Cause	Number	Percent
Stress or Overexertion	45	44.1%
Collision	21	20.6%
Caught or Trapped	16	15.7%
Struck by	9	8.8%
Fell or Jumped	6	5.9%
Assault	3	2.9%
Contact with	2	2.0%
Total	102	100%

Cause of Fatal Injury

As used in this study, the term "cause of injury" refers to the action, lack of action, or circumstances that resulted directly in the fatal injury, while the term nature of injury" refers to the medical cause of the fatal injury or illness, often referred to as the physiological cause of death. A fatal injury usually is the result of a chain of events, the first of which is recorded as the cause. In 2000, a fire chief was assisting with the performance of annual hose testing. A coupling failed during testing and high pressure water struck the chief propelling him back into a fire truck. The chief hit his head on the truck and sustained a fatal head injury. The cause of his fatal injury was recorded as "struck by handline," and the nature of the fatal injury was listed as "trauma," Similarly, if a wildland firefighter was overrun by a fire and died of burns, the cause of the death would be listed as "caught/trapped" by fire progress, and the nature of death would be "burns." This follows the convention used in the NFIRS casualty reports.

Figure 7 shows the distribution of deaths by cause of fatal injury or illness; Table 7 presents the exact number.

Figure 7 - Fatalities by Cause of Fatal Injury (2000)

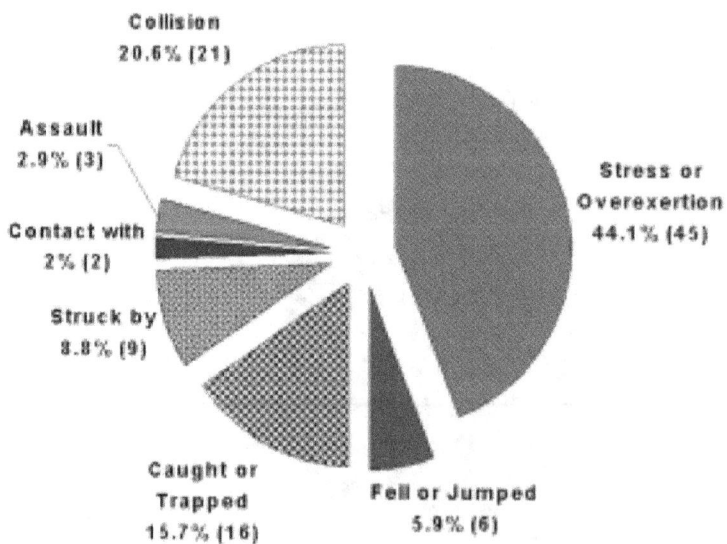

Firefighter Fatalities in the United States in 2000

As in most previous years, the largest cause category is stress or overexertion, which was listed as the primary factor in 44.1 percent of the deaths - the lowest percentage since 1997 when 42.6 of that year's firefighter fatalities were stress-related. Fire fighting is extremely strenuous physical work and is likely one of the most physically demanding activities that the human body performs.

Most firefighter deaths attributed to stress result from heart attacks. Of the 45 stress-related fatalities in 2000, 41 firefighters died of heart attacks, 2 died of CVA's (strokes), and 2 died of other stress-related diseases.

These issues will be explored in the "special topics" section of this report.

The second leading cause of firefighter fatal injuries was collisions. Twenty-one firefighters were killed in collisions in 2000. This number is dramatically higher than the total of 11 for 1999. This is partly attributable to the fact that there were no aircraft-related firefighter fatalities in 1999 and that there were 8 such deaths in 2000.

Two multiple-fatality incidents involving aircraft occurred in 2000, claiming two lives each. One occurred in North Carolina and the other in New Mexico. Aircraft pilots were killed in crashes in Florida, Nevada, and Texas. A firefighter who was a passenger in a helicopter was killed as the aircraft rotated violently upon liftoff and then landed back in its skids.

Thirteen firefighters were killed in 2000 as the result of non-aircraft collisions. Three firefighters were killed in collisions involving their personal vehicles as they responded to emergencies; 3 firefighters that were passengers in responding fire apparatus were killed in collisions; 1 firefighter was killed when the ATV he was driving overturned and he was burned after leaking fuel ignited; 1 firefighter died after his ladder truck collided with a train at a crossing; and 5 firefighters died in separate incidents when the apparatus each was driving was involved in a collision.

The third leading cause of fatal firefighter injuries in 2000 was being caught or trapped, which accounted for 16 deaths.

Table 8.
Deaths Caused by Stress Or Overexertion

Year	Number	Percentage of all Deaths
2000	45	41%
1999	54	49%
1998	42	46%
1997	40	42.6%
1996	47	50.0%

The median age of a firefighter killed as the driver in a collision was 29; the median age for all firefighter deaths was 47.

15

Firefighter Fatalities in the United States in 2000

Table 9.
Deaths Caused by Being
Caught or Trapped

Year	Number	Percentage of all Deaths
2000	16	15.7%
1999	11	10%
1998	16	18%
1997	11	11.7%
1996	7	7.4%

One multiple fatality incident occurred in 2000 where the cause of the fatal injuries was caught or trapped. Two Texas firefighters were trapped in a restaurant fire by the collapse of the building's roof and were killed. Another Texas firefighter was killed when the roof of a church collapsed as he and other firefighters performed ventilation. The firefighter fell into the fire area and could not be rescued in time to save his life.

One firefighter became disoriented in a residential fire and was lost; 1 firefighter fell through a collapsed floor into a basement during a residential fire and was killed; and 1 firefighter was caught in a flashover as he attempted to rescue a woman in an arson-caused apartment building fire.

Seven firefighters were killed when they were caught or trapped by fire progress or sudden fire growth; 4 wildland firefighters were killed when their positions were overrun by fire progress; 2 firefighters were killed when residential structural fires developed quickly and did not allow their escape; and a chief officer was killed during a training burn of an acquired residential structure when fires that he helped to set progressed faster than expected and trapped him in the second story of the structure.

Three drownings in separate incidents in Colorado, Indiana, and North Carolina are also included in the total. In each of these cases, the firefighter is considered as being caught or trapped under water.

Being struck by an object was the fourth leading cause of fatal firefighter injuries in 2000. There were 9 deaths in this category, including 4 firefighters who were struck as they worked at the scene of vehicle collisions and 1 firefighter who was struck by a non-fire department vehicle as he handled a supply line. One firefighter was killed when he was struck by an ARFF vehicle as he tested the pump on his ARFF vehicle; 1 firefighter was struck and killed by a lead plug that had been used to secure a prop weapon used in SWAT-medic training when the weapon was fired; 1 firefighter was killed when he was struck by the top of a 12,000 gallon fuel tank that exploded as it was being cut apart for salvage; and 1 firefighter was killed when he was struck by high-pressure water from a hoseline that failed as it was being tested. He struck his head when he was propelled backward into a piece of fire apparatus.

Six firefighters died when they fell or jumped to their deaths.

Firefighter Fatalities in the United States in 2000

Two firefighters were killed in separate incidents when they fell after working on fire station electric door openers. One firefighter fell from a fire apparatus as he secured a ladder and was injured; he was admitted to and later discharged from the hospital for a back injury. He died of pneumonia the night he was discharged. One firefighter tripped over an object as he investigated the cause of a structure fire and sustained an internal injury; the injury was not diagnosed correctly, and he later died. One firefighter died when his parachute failed to operate during a smokejumper training exercise; and 1 firefighter was killed when he fell and was run over by a ladder truck that he was chasing as it began its response.

Three firefighters were killed in assaults. Two Tennessee firefighters were killed as they arrived on the scene of a fire in a single-family residence. The fire had been intentionally set, and the arsonist killed both firefighters with shotgun blasts as they arrived. One firefighter was killed while participating in a "fill the boot" fund collection drive. A car operated by a gunman pulled up close to the firefighter and the gunman shot the firefighter multiple times.

Two firefighters were killed when they came into contact with extreme weather. The firefighters were part of a wildland team that had been dispatched to assist with a fire. A sudden storm developed and both firefighters were struck by lightning as they took shelter from the storm.

Firefighter Fatalities in the United States in 2000

Figure 8 - Fatalities by Nature of Fatal Injury (2000)

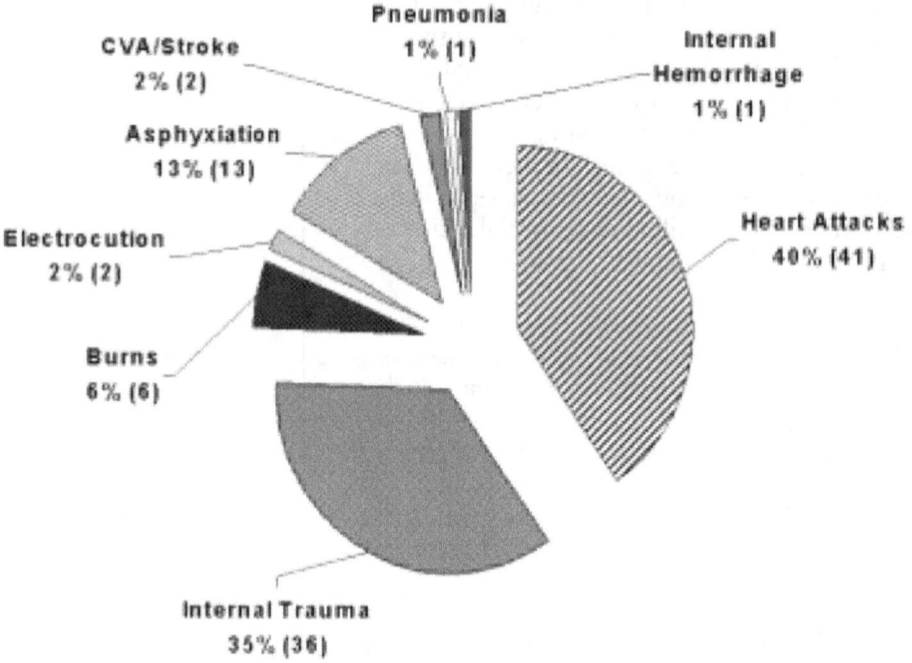

Table 10.
Nature of Fatal Injury - 2000

Nature	Number	Percent
Heart Attack	41	40%
Internal Trauma	36	35%
Asphyxiation	13	13%
Burns	6	6%
CVA/Stroke	2	2%
Electrocution	2	2%
Internal Hemorrhage	1	1%
Pneumonia	1	1%
Total	102	100%

Nature of Fatal Injury

Table 10 and Figure 8 show the distribution of the 102 deaths by the medical nature of the fatal injury or illness. The leading nature of death in 2000 was heart attacks, which accounted for 41 firefighter fatalities.

Figure 9 provides a detailed breakdown of heart attacks by type of duty. Fourteen of the heart attacks occurred at the fire scene, 6 occurred at non-fire scenes, and 6 occurred while responding to or returning from an emergency incident. Seven occurred at training incidents, up sharply from the single such event in 1999 but back to the same level as experienced in 1998. Seven heart attacks occurred during other on-duty situations, and 1 occurred after an incident.

Firefighter Fatalities in the United States in 2000

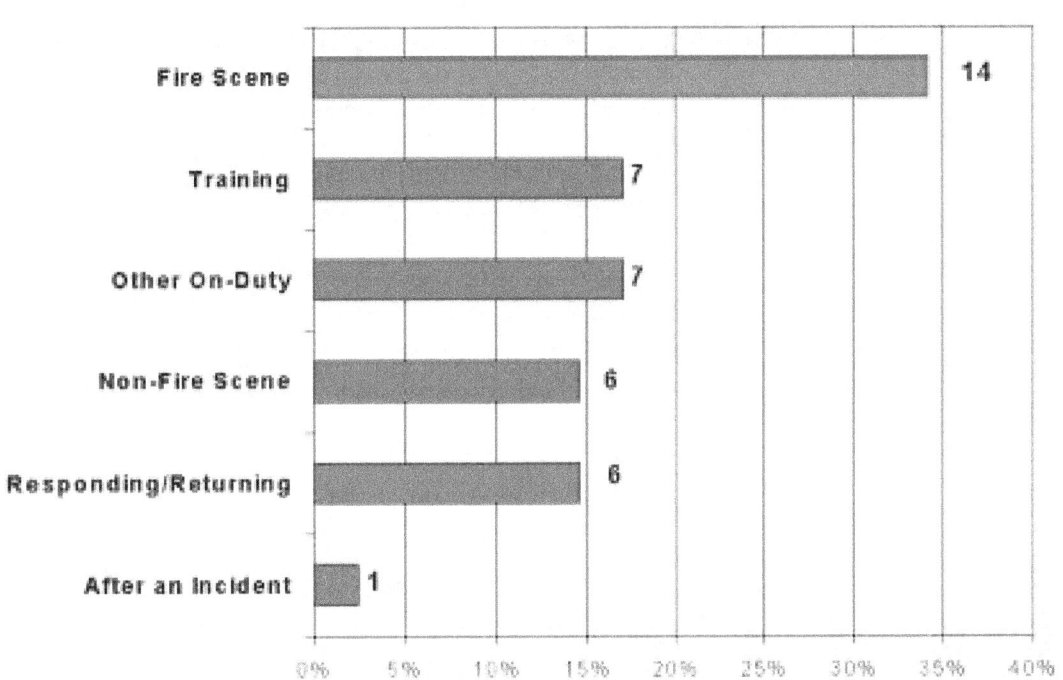

Figure 9 - Heart Attacks by Type of Duty (2000)

Internal trauma was the second leading nature of death, responsible for 36 deaths. This total includes 11 firefighters killed in fire apparatus collisions; 8 firefighters killed in wildland aircraft crashes; 7 firefighters struck by civilian vehicles and fire apparatus; 4 firefighters killed by gunshot wounds; 2 firefighters fell from atop fire apparatus while repairing door closers; 1 firefighter suffered a parachute failure; 1 firefighter fell at the scene of a fire he was investigating; 1 firefighter struck by a piece of a large fuel tank as it was salvaged; and 1 firefighter struck by a hose that failed during testing and suffered a fatal head injury.

Asphyxiation was the third leading medical reason for firefighter deaths, responsible for 13 deaths (see Table 12). Three firefighters drowned in 2000. Two Texas firefighters who became trapped in a restaurant fire died of asphyxiation. Other structural fire fighting asphyxiation deaths include 2 additional Texas firefighters who died in separate incidents, and 1 firefighter each from Delaware, Florida, Michigan, and Utah. Two other firefighters died of asphyxiation in 2000: 1 was a passenger in a fire truck that was involved in a collision and was trapped, and 1 died of acute hypoxia due to pulmonary edema while on the scene of a fire incident.

Table 11.
Internal Trauma
Firefighter Deaths

Year	Number of Firefighter Deaths
2000	36
1999	25
1998	27
1997	32
1996	32

Table 12.
Firefighter Deaths
Due to Asphyxiation

Year	Number of Firefighter Deaths
2000	13
1999	16
1998	15
1997	15
1996	5

19

Firefighter Fatalities in the United States in 2000

Firefighter's Ages

Figure 10 shows the distribution of firefighter deaths by age and nature of the fatal injury. Younger firefighters were more likely to have died as a result of traumatic injuries such as injuries from an apparatus accident or after becoming caught or trapped during fire fighting operations. Stress was shown to play an increasing role in firefighter deaths as age increased.

Figure 10 - Fatalities by Age and Nature

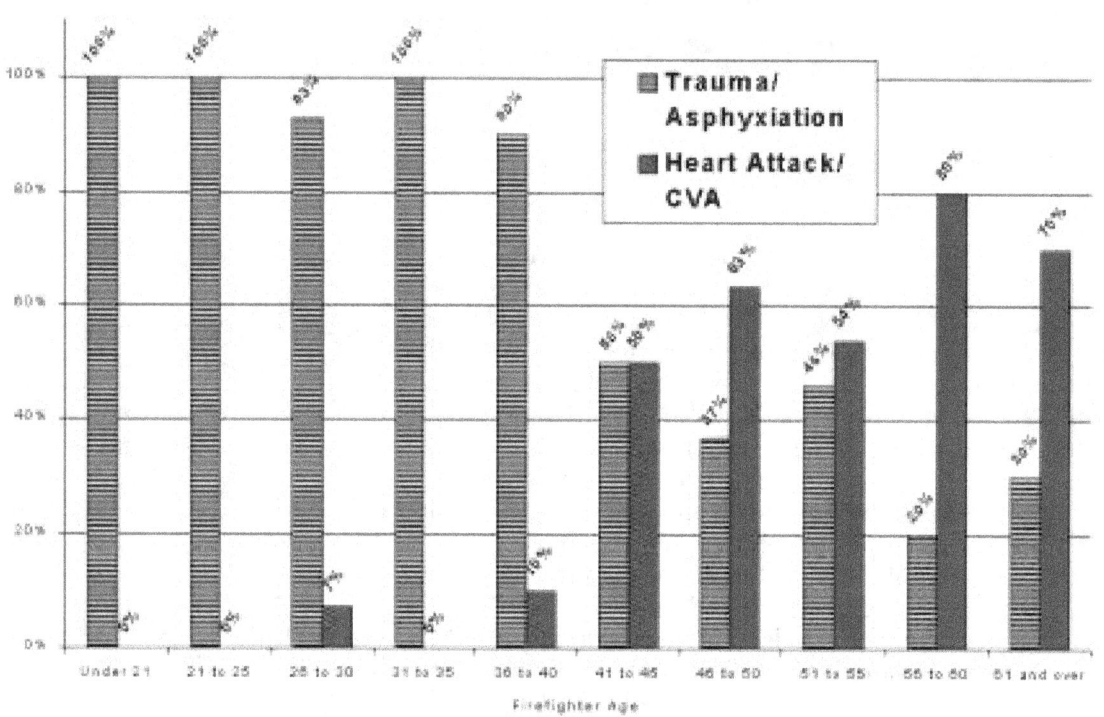

The median age for firefighters who suffered fatal heart attacks or CVAs on-duty in 2000 was 52. Firefighters that died from heart attacks and CVAs ranged in age from 27 to 80. As can be seen from Figure 10, a firefighter between the ages of 41 and 45 that died on-duty in 2000 had a 50-50 chance of dying from a heart attack or CVA versus a traumatic injury or asphyxiation. Older firefighters had a greater chance of dying from a heart attack or CVA than a traumatic injury or asphyxiation.

Firefighter Fatalities in the United States in 2000

Fixed Property Use for Structural Fire Fighting Deaths

There were 25 firefighter fatalities in 2000 where the firefighter became ill while on the scene or engaged in structural fire fighting. Table 13 shows the distribution of these deaths by fixed property use. As in most years, residential occupancies accounted for the highest number of these fireground fatalities, with 21 deaths. Table 14 shows the number of firefighter deaths in residential occupancies for the last 5 years. Residential occupancies usually account for 70 to 80 percent of all structure fires and a similar percentage of the civilian fire deaths each year[1]. Historically, the frequency of firefighter deaths in relation to the number of fires is much higher for non-residential structures.

Type of Activity

Table 15 and Figure 11 show the types of fireground activities the 41 firefighters were engaged in at the time they sus-

Table 13.
Structural Fire Fighting Deaths By Fixed Property Use

Fixed Property Use	Number	Percent
Residential	21	84%
Commercial	3	12%
Manufacturing	1	4%

Table 14.
Firefighter Deaths in Residential Occupancies

Year	Number of Firefighter Deaths
2000	21
1999	23
1998	17
1997	16
1996	19

Figure 11 - Fatalities by Type of Activity (2000)

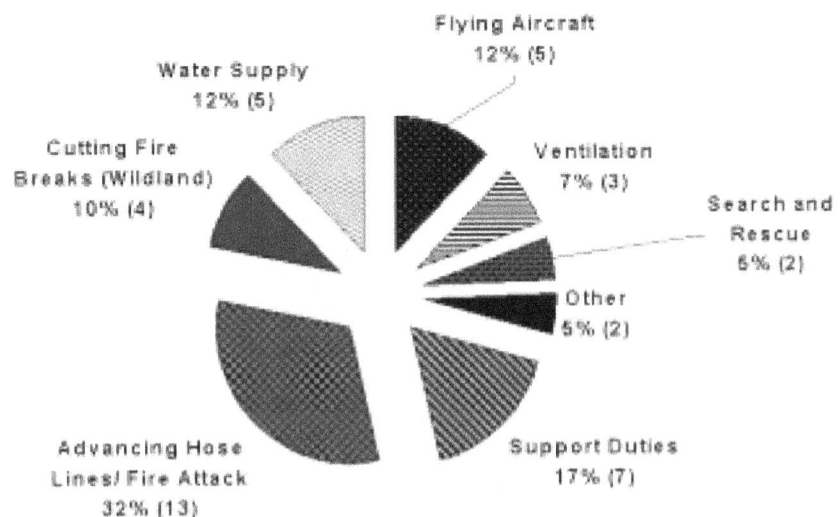

Note: 41 of 102

---

[1] Complete NFIRS fire incidents data were not available at the time of this report, but residential fires typically account for between 70 and 80 percent of all civilian fatalities each year.

tained their fatal injuries or illnesses. This total includes all fire fighting duties such as wildland fire fighting and structural fire fighting.

Thirteen firefighters died while engaged in fire attack and advancing hoselines. Eleven of the 13 firefighters killed were engaged in structural fire fighting; 1 died at a wildland fire; and 1 died extinguishing a small grass fire caused by salvage work on an old fuel tank. The fuel tank later exploded and flying debris killed the firefighter.

Seven firefighters died while engaged in support duties. Two Tennessee firefighters were killed as they arrived on the scene of a residential fire which was deliberately set. The man who set the fire killed both firefighters with shotgun blasts before they could begin fire fighting duties. One firefighter suffered a heart attack as he performed accountability duties; 1 firefighter became ill while photographing a fire scene and later died; 1 firefighter was killed when the ATV he was driving rolled over, fuel was spilled on him, and the fuel ignited. A firefighter was killed when he was struck by a car operated by a drunk driver after the firefighter attached a supply line to a fire hydrant. A firefighter died the night he was discharged from a hospital stay caused by an injury received when he fell from a piece of fire apparatus.

Five firefighters were killed in 4 separate aircraft crashes while conducting fire fighting duties.

Five firefighters were killed while they engaged in water supply tasks. Two wildland firefighters died in separate incidents when their positions were overrun by fire progress; 2 firefighters experienced heart attacks while engaged in pump operations; and 1 firefighter died when he experienced a heart attack as he set up a tanker (tender) fill site to support a structural fire fight.

Four firefighters died while cutting fire breaks. Two Utah firefighters were struck by lightning and killed as they took shelter from a storm under some trees; 1 Mississippi firefighter was killed when his bulldozer was overrun by fire progress as he attempted to contain a fire; and a Wyoming firefighter was killed when a pile of wood debris that was thought to be controlled "blew up" and severely burned him.

Table 15.
Type of Activity - 2000

Nature	Number	Percent
Advancing Hoselines/ Fire Attack	13	32%
Support Duties	7	17%
Flying Aircraft	5	12%
Water Supply	5	12%
Cutting Fire Breaks (Wildland)	4	10%
Ventilation	3	7%
Search and Rescue	2	5%
Other	2	5%
Total	41	100%

Table 16.
Firefighter Deaths in Fire Attack/Advancing Hoselines

Year	Number of Firefighter Deaths
2000	13
1999	16
1998	18
1997	21
1996	9

Firefighter Fatalities in the United States in 2000

Ventilation was the duty being performed by 3 firefighters who died in 2000. A Texas firefighter fell through the roof of a fire-involved church and perished; and firefighters in Illinois and Iowa died of heart attacks while performing ventilation on residential structures to search for fire extension.

Two firefighters were killed as they conducted search and rescue activities. A Florida firefighter died searching a residence, and a Michigan firefighter died while trying to remove a resident from an apartment building that was on fire. Both firefighters were caught by fire progress.

Two firefighters died while performing duties that are not classified in any other category. A firefighter in North Carolina died of hypoxia, not related to smoke exposure, while performing support tasks at a fire scene; and a Kentucky firefighter suffered a CVA (stroke) while searching for the source of an automatic fire alarm in a nursing home.

Time of Injury

The distribution of all 2000 firefighter deaths according to the time of day when the fatal injury occurred is illustrated in Figure 12 (one incident time was unable to be determined).

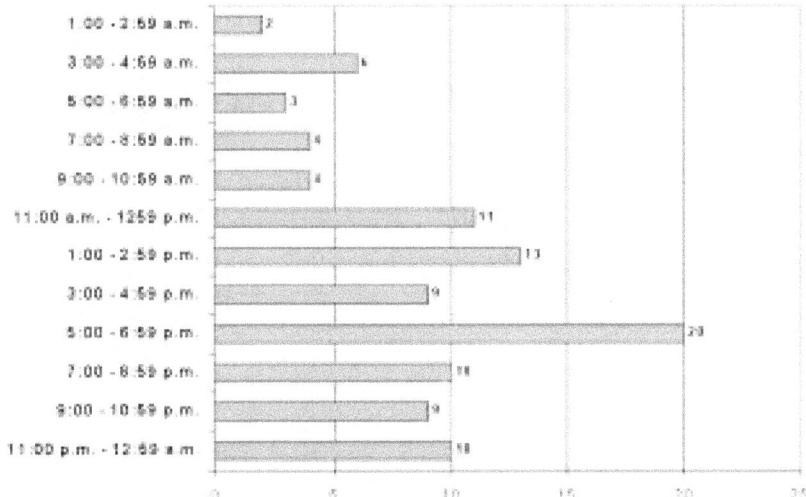

Figure 12 - Fatalities by Time of Fatal Injury (2000)

Note – 101 of 102, the time of injury for one death was unable to be determined.

Firefighter Fatalities in the United States in 2000

Month of the Year

Figure 13 illustrates firefighter fatalities by month of the year. Firefighter fatalities peaked in August with 15 deaths. Other than the fact that wildland fires and their associated injuries and deaths occur in the wildland season, no trends were identified.

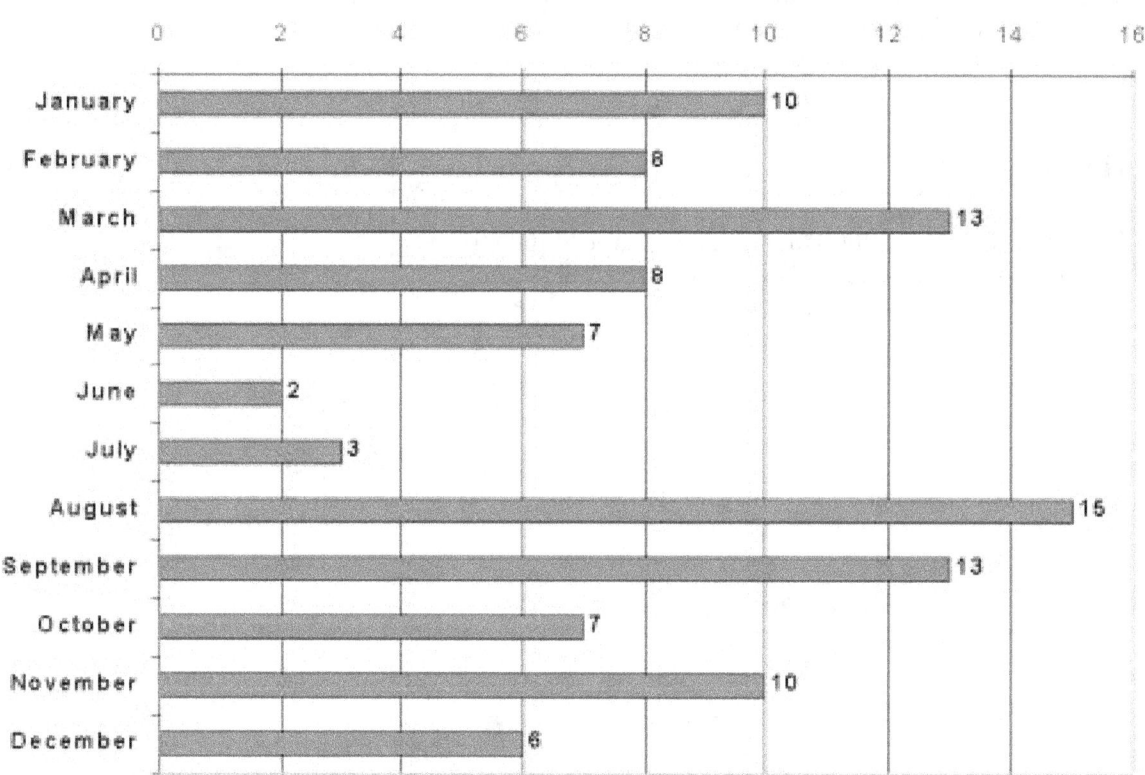

Figure 13 - Deaths by Month of the Year (2000)

Firefighter Fatalities in the United States in 2000

State and Region

The distribution of firefighter deaths by State is shown in Table 17.[2] Thirty-nine States and Puerto Rico each had at least 1 firefighter fatality. Texas had the highest number of deaths with 11 followed by Pennsylvania with 8. Figure 14 shows the firefighter fatalities divided by region of the country and their status as career, volunteer, or wildland firefighters.

Figure 14 also provides information on the ratio of firefighter fatalities per million population in each region. The Northeast and South had higher firefighter death rates than the Northcentral or West.

Figure 14.
Firefighter Deaths by Region (2000)

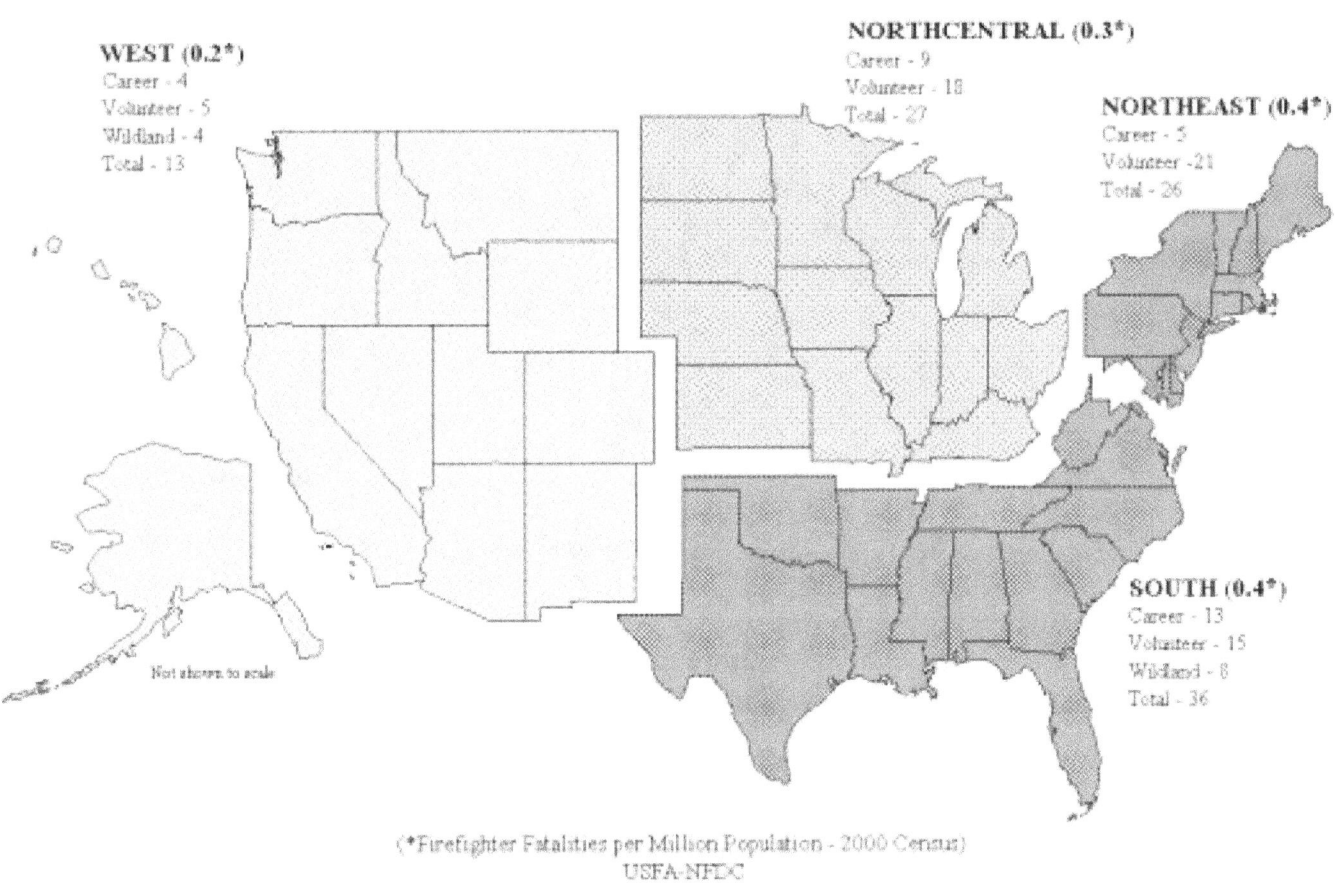

WEST (0.2*)
Career - 4
Volunteer - 5
Wildland - 4
Total - 13

NORTHCENTRAL (0.3*)
Career - 9
Volunteer - 18
Total - 27

NORTHEAST (0.4*)
Career - 5
Volunteer - 21
Total - 26

SOUTH (0.4*)
Career - 13
Volunteer - 15
Wildland - 8
Total - 36

Not shown to scale

(*Firefighter Fatalities per Million Population - 2000 Census)
USFA-NFDC

---

[2] This list attributes the deaths according to the State in which the fire fepartment or unit is based, as opposed to the state in which the death occurred. They are listed by those States for statistical purposes, and for the National Fallen Fiefighters memorial at the National Emergency Training Center.

Firefighter Fatalities in the United States in 2000

The figure below shows the location of each firefighter fatality for 2000.

Analysis of Urban/Rural/Suburban Patterns in Firefighter Fatalities

The United States Bureau of the Census defines "urban" as a place having a population of at least 2,500 or lying within a designated urban area. Rural is defined as any community that is not urban. Suburban is not a census term but may be taken to refer to any place, urban or rural, that lies within a metropolitan area defined by the Census Bureau, but not within one of the central cities of that metropolitan area.

Firefighter Fatalities in the United States in 2000

Fire department areas of responsibility do not always conform to the boundaries used for the census. For example, fire departments organized by counties or special fire protection districts may have both urban and rural coverage areas. In such cases, it may not be possible to characterize the entire coverage area of the fire department as rural or urban, and firefighter deaths were listed as urban or rural based on the particular community or location in which the fatality occurred.

The following patterns were found for 2000 firefighter fatalities. These statistics are based on answers from the fire departments and, when no data from the department were available, the data are based upon population and area served reported by the fire departments.

Table 18.
Firefighter Deaths by Coverage Area Type

	Urban/ Suburban	Rural	Federal or State Parks/Wildland	Total
Firefighter Deaths	48	42	12	102

Table 17.
2000 - States with On-Duty Firefighter Fatalities

State	Number of Deaths
Alabama	2
Alaska	1
Arizona	1
Arkansas	1
California	4
Colorado	1
Connecticut	4
Delaware	1
Florida	3
Georgia	2
Illinois	5
Indiana	2
Iowa	3
Kansas	1
Kentucky	4
Louisiana	1
Maine	1
Maryland	1
Massachusetts	1
Michigan	3
Mississippi	1
Missouri	5
Nevada	1
New Hampshire	1
New Jersey	3
New Mexico	1
New York	5
North Carolina	6
Ohio	3
Oklahoma	4
Pennsylvania	8
South Dakota	1
Tennessee	2
Texas	11
Utah	3
Vermont	1
Virginia	1
West Virginia	1
Wyoming	1
Puerto Rico	1
Total	**102**

Firefighter Fatalities in the United States in 2000

CONCLUSIONS

The year 2000 was another tragic year of loss for America's firefighters. The 102 lives lost in 2000 is a welcome reduction from the level of 112 that occurred in 1999. However, this number is still not low enough.

Violence against firefighters returned in 2000 after a lull in such incidents. Gun shot homicides while on-duty killed three firefighters. This fact points to the need for continued caution.

Another type of violence, arson fires, claimed 10 firefighters in 2000. A firefighter killed while fighting an arson fire is considered a homicide in most situations.

After a year with no such deaths, wildland aircraft crashes claimed the lives of 8 firefighters in 2000.

The five-year trend for firefighter fatalities is an increase of 10 percent. The USFA has set two ambitious goals for reductions in firefighter fatalities:

- A 25 percent reduction in on-duty firefighter fatalities within 5 years.

- A 50 percent reduction in on-duty firefighter fatalities within 10 years.

The perennial killer of firefighters returned in 2000; heart attacks claimed the largest portion of firefighters, 41 in the year 2000 alone. Over the last 5 years, heart attacks have killed 213 firefighters. There is no quick solution to this problem but there is much more that the fire service can do to reduce the incidence of heart-related firefighter deaths. This report explores the causes of heart attacks, discusses risk factors, and provides information about reducing the impact or eliminating heart disease risks.

This report poses some immediate impact ideas to help save the lives of firefighters in the future. Each year, firefighters are killed in situations that could have easily been prevented. The death could have been avoided or the severity of the injury could have been minimized through some free or inexpensive action taken prior to the emergency. The use of seat belts, PASS devices, prudent driving, and EMS standby at incidents and training exercises can have an immediate impact in lives saved.

Firefighter Fatalities in the United States in 2000

SPECIAL TOPICS

HEART ATTACKS

Two-hundred-thirteen on-duty firefighters died of heart attacks in the five-year period beginning on January 1, 1996 and ending December 31, 2000.

Heart attacks kill over 4 out of every 10 firefighters that die on-duty. In the past 5 years, the 213 firefighter deaths due to heart attacks represent 43 percent of the 494 firefighter deaths that occurred in the period.

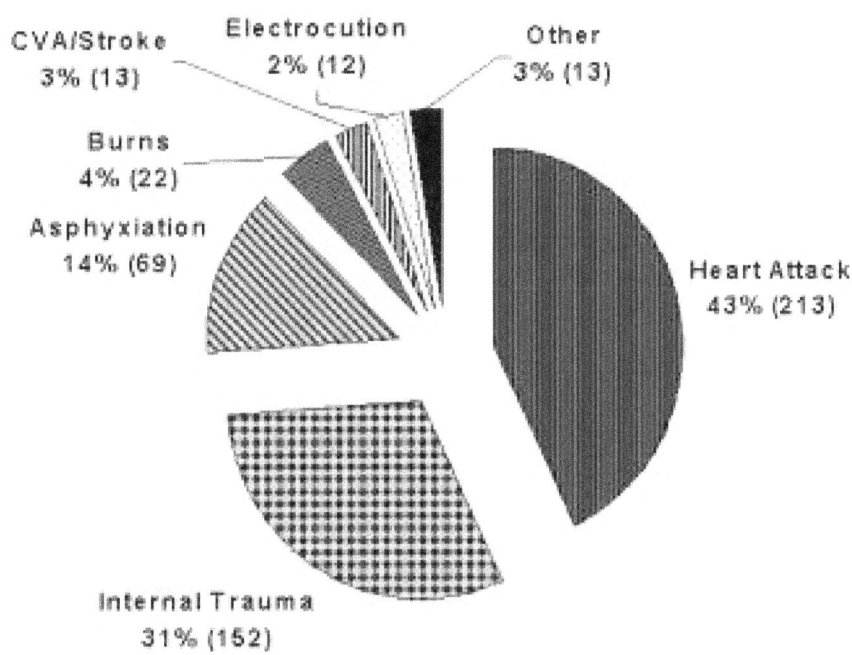

Nature of Fatal Injury 1996-2000

Firefighter Fatalities in the United States in 2000

In 2000, thirteen firefighters who died of heart attacks had a previous history of heart disease, heart attacks, or bypass surgery.

The firefighters that died of heart attacks in the last 5 years ranged in age from 24 to 90. The average age of a firefighter who died on-duty from a heart attack was 53 years, 1 month, and 28 days.

Heart attacks are the largest single killer of firefighters. From 1996 through 2000, heart attacks claimed nearly as many firefighters as all of the deaths from internal trauma and asphyxiation combined (there were a total of 221 internal trauma and asphyxiation deaths).

Fatal Heart Attacks and Firefighters

Over one-half of all fatal heart attacks that strike firefighters do so on the scene of a fire or non-fire emergency. Fire fighting activity and work on the scene of many non-fire emergencies are extremely strenuous work. For this reason, the cause of the vast majority of firefighter fatalities due to heart attacks is classified as stress or overexertion. The physiological stress that is placed on the firefighter's body during these activities causes the heart attack.

Fatal Heart Attacks by Type of Duty 1996-2000

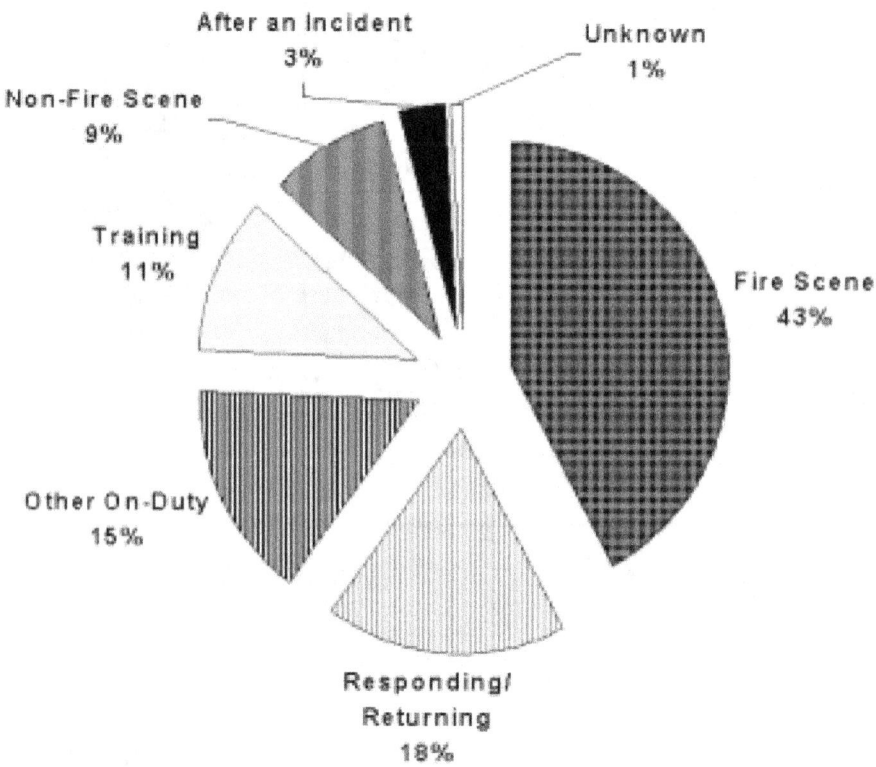

Firefighter Fatalities in the United States in 2000

From 1996 through 2000, 73 percent of firefighter heart attack deaths were related to emergency duty. This number includes firefighters who suffered heart attacks while responding to or returning from an incident, at an emergency scene, or just after the conclusion of an emergency.

Career and Volunteer Heart Attack Deaths

Heart attacks strike both volunteer and career firefighters, 146 volunteers and 67 career firefighters died of heart attacks from 1996 through 2000. These numbers represent 46 percent of all on-duty volunteer firefighter deaths and 38 percent of all on-duty career firefighter deaths that occurred in that time period.

What are Risk Factors Associated with Heart Disease?

According to the American Heart Association (AHA), there are a number of risk factors that place an individual at higher risk of suffering a heart attack. These risk factors are:

Tobacco Smoke

For the overall population, about 1 in 5 deaths that occur from cardiovascular (heart) diseases are attributable to smoking. Smoking among American adults has declined by approximately 42 percent since 1965. However, this downward trend may have recently leveled off. Non-smokers are also at risk from second-hand smoke.

Cholesterol and Other Lipids

High levels of blood cholesterol lead to the development of fatty deposits on the interior of blood vessels, including the critical blood vessels that provide the heart with blood. These deposits can build up and limit blood flow, thereby decreasing the work that the heart can perform. Blood clots can be created when these deposits break loose and can block blood flow to the heart causing a heart attack.

Physical Inactivity

The relative risk of heart disease presented by a lack of physical activity is roughly the same as the risk posed by high blood cholesterol, high blood pressure, or smoking. Less active, less physically fit persons have a 30 to 50 percent higher chance of developing high blood pressure than their active, fit counterparts.

Weight

A large number of Americans are considered overweight or obese, 150 million in all. Excess body fat and weight place a strain on the heart.

Diabetes Mellitus

Diabetes is an inability of the body to process sugars into energy. This disease has been diagnosed in over 10 million Americans. An estimated 5 million Americans have diabetes and do not know it since it has not been diagnosed by a medical professional. Two-thirds of people with diabetes die of some form of heart or blood vessel disease. The majority of people with diabetes can treat the disease through diet or medication.

The Firefighter Fatality Investigation and Prevention Program, a component of the National Institute for Occupational Safety and Health (NIOSH) provides the following additional risk factors:

Age, Gender, and Family History

Males over age 45 are at higher risk for Coronary Artery Disease (CAD). Persons with a family history of CAD are also at higher risk.

High Blood Pressure

High blood pressure has been found to be a risk factor for heart disease.

How Can Heart Attacks in Firefighters be Prevented?

There is not much that can be done to address some of the risk factors listed above. Advancing age, gender, and family history are factors that cannot be impacted. There are, however, a number of risk factors that can be addressed and minimized.

The **American Heart Association**
The primary source of information on risk factors, heart disease, and the prevention of heart attacks is the AHA. The AHA sponsors a very complete internet site at http://www.americanheart.org.

The AHA provides an internet-based health program entitled "One of a Kind." Participation in the program is simple. The program starts with the completion of an on-line survey that takes approximately 15 minutes. The answers that you provide in the survey help the AHA select the types of materials that will help you achieve your health goals. You can visit the AHA web site to review your progress at any time and link to information tailored to your situation and goals. The system also generates e-mails that remind you of your goals and provide encouragement. The program is free to the user and all information is kept confidential.

Information is also available from local AHA offices and through the mail from their National Center. Additional information about the AHA is available at the end of this section.

Medical Evaluations and Examinations

The key to the discovery and treatment of physical problems and lifestyle issues that may lead to illness is a physical evaluation or examination. These procedures can identify risk factors and diseases that may be present in the firefighter. Once these risk factors and health concerns are known, they can be treated, and the level of risk for sudden cardiac death and other causes of death can be reduced.

Annual medical evaluations are required by NFPA 1582, *Standard on Medical Requirements for Fire Fighters and Information for Fire Department Physicians*. The medical evaluation consists of a medical history, an occupational and exposure history, measurements of height and weight, a blood

> *In 2000, 32 of the 102 firefighters that died had never been given a department sponsored medical exam.*

pressure check, and a heart rate and rhythm check. The annual evaluation does not necessarily need to be performed by a physician. The standard allows the evaluation to be performed by a qualified person and reviewed by a physician.

A more thorough medical check is also required by the standard, although the frequency is less than annually for younger firefighters. The frequency of this check is dependent upon the firefighter's age. A medical examination is required at least every 3 years for firefighters aged 29 and under; at least every 2 years for firefighters aged 30 to 39; and every year for firefighters aged 40 and above. A physician must perform the medical examination.

The NFPA 1582 standard also requires that medical evaluations and examinations be performed at no cost to the fire department member. Medical evaluations are also required prior to becoming a firefighter and prior to returning to duty after an extended absence due to illness or other reasons.

Smoking

NFPA 1500, *Standard on Fire Department Occupational Safety and Health Program* requires that fire departments provide their members with information on the hazards of tobacco use, including smoking. The standard also requires that the fire department provide a smoking cessation program to its members.

Advances in medications used to help smokers quit have allowed many people to stop the habit. Nonetheless, nicotine dependency is a very difficult problem to overcome. Help is available from two well-know organizations: the American Cancer Society and the American Lung Association. Contact information for both organizations appear at the end of this section.

A program entitled the Quit Smoking Action Plan is available from the American Lung Association through the internet. The internet sites of both organizations offer a wealth of information on smoking cessation programs, the benefits of smoking cessation, and studies on the effectiveness of different approaches. Use the site search features and search on the term "smoking cessation."

The good news is that studies have shown that an individual who quits smoking cuts his/her risk of heart disease in half after 1 year of non-smoking. Former smokers that have stopped smoking have approximately the same risk of heart disease as a non-smoker after 15 years smoke free.

Cholesterol

Everyone should know their cholesterol numbers. This test has become an inexpensive and standard part of any annual physical assessment by a medical professional. Total blood cholesterol levels above 200 milligrams per deciliter (mg/dl) are considered borderline or high risk. Levels above 240 mg/dl are considered high risk. A 10 percent reduction in blood cholesterol levels has been shown to reduce the chances of coronary heart disease by 30 percent.

A person with a cholesterol level of 240 mg/dl is twice as likely to have a heart attack as a person with a cholesterol level of 200 mg/dl. You can reduce the levels of cholesterol in your blood by eating a low fat diet, losing weight if you need to, and exercising 30 to 40 minutes 3 times a week. Medications are also available from your physician that can help to lower your blood cholesterol. Additional information on blood cholesterol and ways to reduce blood cholesterol can be obtained from the AHA. A very useful and informative section of the AHA web site that deals with blood cholesterol can be found at http:// www.americanheart.org/cholesterol/do.jsp

Physical Inactivity and Weight

NFPA 1500 requires fire departments to provide a physical fitness program for their members. The standard also requires the structured participation of all fire department members in the physical fitness program. NFPA 1583, *Standard on Health-Related Fitness Programs for Fire Fighters*, provides minimum requirements and information on the components of a physical fitness program for firefighters.

The *Fire Service Joint Labor Management Wellness-Fitness Initiative,* which was developed by the International Association of Fire Chiefs (IAFC) and the International Association of Fire Fighters (IAFF), provides a framework for long-term attention to issues of firefighter health and wellness. Copies of this report are available for purchase from the IAFC. Copies are also available to IAFF members from the IAFF.

Diabetes Mellitus and High Blood Pressure

Like blood cholesterol, many cases of high blood pressure and diabetes can be controlled with diet, exercise, and medication. Information on diabetes may be obtained from the American Diabetes Association. Information on high blood pressure may be obtained from the AHA.

Signs of a Heart Attack and the Importance of Immediate Action

In several instances in 2000, firefighters became ill and exhibited signs of a heart attack. These symptoms were minimized by the firefighter, treatment was ignored or refused, and the opportunity for early heart attack care was missed. Firefighters should be alert for signs of a heart attack in themselves and in those who work with them. Immediate action, including talking a reluctant coworker into a hospital emergency room visit, could save a life.

Early signs of a heart attack are familiar to many firefighters since they are detailed in first responder and emergency medical technician training. Early signs of a heart attack include profuse sweating, chest pain, weakness, nausea, dizziness, and chest pressure. These signs may present themselves at the same time or they may not be present at all.

If any of these symptoms are present, a firefighter should seek immediate medical care. If a firefighter sees these symptoms in another firefighter, he or she should do everything possible to convince the sick firefighter to seek aid.

An excellent resource with information on early heart attack signs and advice on convincing someone to seek medical aid in these situations is available in a brochure from Saint Agnes HealthCare in Baltimore. The program is The Early Heart Attack Care Partnership (EHAC). Contact information is at the end of this section.

Sources of Additional Information

American Cancer Society
800-ACS-2345
www.cancer.org

American Diabetes Association
1701 North Beauregard Street
Alexandria, VA 22311
800-342-2383
www.diabetes.org

American Heart Association
National Center
7272 Greenville Avenue
Dallas, TX 75231
800-AHA-USA1
www.americanheart.org

American Lung Association
1840 Broadway
New York, NY 10019
800-586-4872
www.lungsusa.org

HEALTHY PEOPLE 2010 – Healthy People in Healthy Communities
United States Department of Health and Human Services
Office of Disease Prevention and Health Promotion
Room 738G Hubert Humphrey Building
200 Independence Avenue, SW
Washington, DC 20201
202-205-8583
www.health.gov/healthypeople/default.htm

International Association of Fire Chiefs
4025 Fair Ridge Drive
Fairfax, VA 22033-2868
www.iafc.org

Firefighter Fatalities in the United States in 2000

International Association of Fire Fighters
1750 New York Avenue, NW
Washington, DC 20006
www.iaff.org

The Early Heart Attack Care Partnership
Saint Agnes HealthCare
The Paul Dudley White Coronary Care System
Baltimore, MD 21229
410-368-3200
www.ehac.org

IMMEDIATE IMPACT AREAS

Every year, a number of firefighters are killed in situations that could have been prevented through some simple action or through the provision of an inexpensive piece of equipment. The year 2000 was no different than previous years. This section will explore some inexpensive ways to make an immediate impact on firefighter safety.

The following list of immediate impact ideas most certainly would have saved the lives of firefighters in 2000, and these same ideas can certainly save the lives of firefighters in the future.

Most of the ideas proposed in this section can be done without cost.

IMMEDIATE IMPACT AREA #1 – SEAT BELT USE

Seven potential saves in 2000

Each year firefighters are needlessly killed in collisions that would have been survivable if the firefighter had been wearing a seat belt. In most cases, seat belts were present in the vehicle but were not in use.

Seat belts cannot guarantee survival in a collision: 3 firefighters that were wearing seat belts died in collisions in 2000.

Firefighter Fatalities in the United States in 2000

Seat belts do give a firefighter a better chance of survival and minimize the chance that the firefighter will be ejected from the vehicle. In 2000, 7 firefighters that were not wearing seat belts were killed in collisions, a number of others were injured.

Fire department members should:

- Make sure that seat belts are present and operable in all fire department vehicles.

- Adopt mandatory seat belt use policies.

- Educate fire department members on the benefits of seat belt use and the consequences that may occur if they are not used.

- Place signs and placards that can be viewed by firefighters seated in fire department vehicles that provide a reminder of the policy on seat belt use.

IMMEDIATE IMPACT AREA #2 – PASS DEVICE USE

Two potential saves in 2000

A PASS device is worn by a firefighter either attached to the firefighter's protective clothing or attached or integrated into the firefighter's SCBA. It is designed to emit a loud audible tone if the device does not sense that the firefighter is moving. The sound is intended to both alert others on the fireground that a firefighter is in danger and to assist rescuers in locating the downed firefighter.

Many firefighters have been equipped with PASS devices; their use has been required by NFPA 1500 for over a decade. Some newer PASS devices activate automatically when the firefighter readies the Self-Contained Breathing Apparatus (SCBA) for service; the PASS switches on when air from the cylinder is released into the SCBA. Other PASS devices are not connected to the SCBA and must be manually activated by the firefighter; the firefighter has to turn a knob or press a button(s) to put the PASS device in-service.

Each year firefighters are lost or trapped in structures while wearing manually activated PASS devices in the "off" position. In at least two cases in 2000, firefighters who were lost or trapped in burning structures were wearing manually activated PASS devices that were off. Even though they knew that their lives were in imminent danger, they failed to activate their PASS devices. This is a continuing problem. Rescuers had difficulty in locating the downed firefighters. A PASS device sounding an alarm might have helped with the search and possibly produced a different outcome for the lost firefighters.

The only ways to assure that a PASS will be activated when it is needed are to either provide PASS devices that activate automatically or to make it a Standard Operating Procedure (SOP) to turn a manually activated PASS device to the "armed" position whenever a firefighter dons an SCBA.

To make sure that PASS devices work to help save firefighters, the following steps should be taken:

- Adopt an SOP that requires firefighters to activate their PASS devices whenever an SCBA is being worn, even if the facepiece has not yet been donned. Train firefighters in the application of the SOP.

- Everyone on an emergency scene should help to assure that PASS devices are activated. Most PASS devices incorporate flashing lights or knobs that show a different color to allow an observer to determine if the PASS is activated from a distance. If a firefighter sees a PASS in the "off" position, he or she should remind the other firefighter of the need to activate the device.

- Incorporate PASS alarm awareness into firefighter survival training. Teach firefighters to sound their PASS device when they are in danger.

- As funding allows, transition to PASS devices that are integrated with the SCBA.

IMMEDIATE IMPACT AREA #3 – SLOW DOWN

Four potential saves in 2000

Vehicle collisions, rollovers, and collisions with other vehicles and objects are responsible for a number of firefighter deaths each year. Many of these collisions result from firefighters who are driving too fast for conditions. Adverse driving conditions include darkness, weather, road configuration, and vehicle configuration. On a 5-mile response, the response time difference between a response speed of 60 mph and 50 mph is 1 minute.

To make sure that firefighters arrive safely at the emergency scene and safely back to the fire station, the following actions should be considered:

- Wear seat belts (mentioned earlier).

- Drive slower rather than faster, especially when returning to quarters.

- Take vehicle configuration into account. Tankers are notorious for presenting control problems, slow down!

- If the vehicle's right wheels leave the paved surface of the road, do not wrestle the vehicle back on the road. Slow down and bring the vehicle back onto the road when it is safe to do so.

IMMEDIATE IMPACT AREA #4 – EMS STANDBY

Five potential saves in 2000

Fire fighting is an extremely hazardous type of work. Training for fire fighting can present hazards similar to actual fire fighting hazards. The presence of trained emergency medical personnel on the scene of fire fighting and training exercises will enhance medical treatment for civilian victims and it will also enhance the survivability of firefighters.

NFPA 1500 requires the incident commander to evaluate the need for standby medical assistance at emergencies. Such assistance is required by the standard for some incidents.

Emergency medical personnel should stand by at all fire fighting and training exercises. Their presence and quick intervention in a medical emergency involving a firefighter could save a life. If the fire department does not have emergency medical capabilities, a local ambulance squad or other pre-hospital provider may agree to be present in these situations.

Firefighter Fatalities in the United States in 2000

APPENDIX A
SUMMARY OF 2000 INCIDENTS

If additional information is available regarding a firefighter fatality, the reader is directed to these sources. Where possible, hyperlinks that direct the reader to additional information are provided. Hyperlinks operate in the digital version of this report and appear in all versions as underlined text. If links have expired or if the reader does not have internet access, contact information for these sources is provided at the end of the appendix when available.

January 8, 2000
Lee A. Purdy, Pump Operator/Inspector
Age 57, Volunteer
Spencerville Invincible Fire Company, Ohio

Pump Operator/Inspector Purdy was operating a top-mounted pump panel at the scene of a residential structure fire. Pump Operator/Inspector Purdy asked his wife, a volunteer paramedic, for a drink. When she returned to the truck with the drink, she saw him fall from the truck, the victim of a massive heart attack. Medical aid was provided immediately. Pump Operator/Inspector Purdy was transported to a local hospital where he was pronounced dead 20 minutes after his arrival. No autopsy was performed.

January 11, 2000
Ronald J. Osadacz, First Assistant Chief
Age 36, Volunteer
Morganville Volunteer Fire Company Number One, New Jersey

First Assistant Chief Osadacz was on the scene of a vehicle fire that resulted from the collision of a pickup truck with a tree. While working on the scene, First Assistant Chief Osadacz was struck in the groin area by a water stream from a 1½-inch hoseline. First Assistant Chief Osadacz was agitated by this occurrence, left the scene, and returned to his home. Upon his arrival at home, First Assistant Chief Osadacz complained of indigestion, took some over-the-counter medicine, and laid down to rest. Within a few moments, First Assistant Chief Osadacz was struck with a fatal heart attack. An autopsy revealed that First Assistant Chief Osadacz died of severe occlusive coronary arteriosclerosis. A physician who examined First Assistant Chief Osadacz less than a week prior to his death stated that the Chief's agitated state would have contributed to the heart attack.

Firefighter Fatalities in the United States in 2000

January 11, 2000
Allen L. Streeter, Firefighter
Age 58, Volunteer
Ranch Drive Fire District, Oklahoma

Firefighter Streeter responded to a trash and grass fire in the department's brush truck. Shortly after exiting the vehicle, Firefighter Streeter collapsed of an apparent heart attack. CPR was administered immediately by other firefighters, and an ambulance was summoned. Firefighter Streeter was a charter member of the Ranch Drive Volunteer Fire Department.

January 12, 2000
Robert M. Jones, Firefighter
Age 48, Volunteer
Unity Volunteer Fire Department, Maine

Firefighter Jones was preparing to use a dry hydrant to supply water in support of a fire fight in a residential structure. Firefighter Jones attached a large diameter hoseline to the pumper, removed the cap from the dry hydrant, and was preparing to attach a suction hose to the hydrant. Another firefighter, who was assisting Firefighter Jones, came around the truck and found Firefighter Jones on the ground. After calling for help, the assisting firefighter began CPR. Despite treatment on the scene and in the ambulance, Firefighter Jones died. The cause of death was listed as a heart attack. Firefighter Jones had served as the Chief of his fire department from 1990-98, and was currently serving as the Sheriff of Waldo County. Firefighter Jones died on his 48th birthday. The fire was caused when one of three triplet boys inadvertently set a sofa on fire. All three boys, who were age 6, died in the fire.

January 15, 2000
Gary Lynn Bankert, Sr., Firefighter
Age 37, Volunteer
Roanoke-Wildwood Volunteer Fire Department, North Carolina

Firefighter Bankert was participating in fire department sponsored dive training in a rock quarry that contains a private lake used exclusively for recreational diving. Firefighter Bankert was a member of his department's search and recovery dive team. As the class ascended from the third of three dives, the class stopped for a safety and accountability check at a depth of 15 feet. At the time of the check, Firefighter Bankert was present; however, when the class proceeded to the surface, Firefighter Bankert did not surface. Other divers went immediately to the bottom of the lake and found Firefighter Bankert at a depth of approximately 22 feet. Firefighter Bankert was brought to the surface and transported by paramedic ambulance to a local hospital. Despite efforts on the scene and at the hospital, Firefighter Bankert was pronounced dead later that evening. The cause of death was listed as severe metabolic acidosis as the result of a near drowning.

Additional information about this incident may be found in NIOSH Fire Fighter Fatality Investigation F2000-11.

Firefighter Fatalities in the United States in 2000

January 16, 2000
Ernest John Young, Firefighter/Trustee
Age 52, Volunteer
Big Knob Volunteer Fire Department — Station 26, Pennsylvania

Firefighter/Trustee Young was assisting with the replacement of electric garage door openers on some apparatus bay doors at his fire station. Firefighter Young and another firefighter had climbed to the top of a fire rescue truck using a 14-foot extension ladder. As Firefighter Young began his descent, the ladder slipped out from under him and Firefighter Young fell approximately 10 feet and struck his head on the concrete floor. The ladder was not being footed at the time it fell. Despite immediate medical aid and transport by helicopter to a regional hospital, Firefighter Young died on January 17, 2000. The cause of death was listed as blunt force trauma to the head.

Additional information about this incident may be found in NIOSH Fire Fighter Fatality Investigation F2000-07.

January 17, 2000
James William Altic, Fire Chief
Age 47, Volunteer
Halfway Fire & Rescue, Missouri

Chief Altic was the lone occupant and driver of a tanker (tender) apparatus responding with lights and siren to a mutual aid structure fire. Road conditions were slippery because a light misty rain was falling after a long period without rain. Chief Altic failed to negotiate a curve in the road, and the apparatus left the roadway and rolled over. Chief Altic sustained fatal neck and chest injuries and was pronounced dead at the scene. The driver's seat belt had been removed from the apparatus at some point prior to the collision, so Chief Altic was not equipped with a seat belt. The police report cited the speed of the fire apparatus as a factor in the collision, as well as the wet roadway.

Additional information about this incident may be found in NIOSH Fire Fighter Fatality Investigation F2000-18.

Firefighter Fatalities in the United States in 2000

January 17, 2000
Juan Gilberto De Leon, Captain
Age 53, Career
McAllen Fire Department, Texas

Captain De Leon was in his assigned sector driving a command vehicle. He stopped at a business in his sector to help a civilian move some boxes. During this task, Captain De Leon was struck with a heart attack. The civilian activated the 9-1-1 system and provided CPR until the arrival of fire department and EMS responders. The cause of death for Captain De Leon was listed as a myocardial infarction (heart attack).

Additional information about this incident may be found in NIOSH Fire Fighter Fatality Investigation F2000-12.

January 27, 2000
Walter Harvey Gass, Captain
Age 74, Volunteer
Sealy Volunteer Fire Department, Texas

Captain Gass and other members of his department were dispatched to a residential structure fire that was caused when lightning struck a house. The first two firefighters on the scene, the Assistant Chief and the Fire Chief, confirmed a working fire with dark smoke and fire visible from the attic and dormers. Captain Gass and his crew were the first fire company to arrive at the scene. Captain Gass and two firefighters entered the structure through the front door to perform an aggressive attack on the fire. Shortly after entering the structure, the two firefighters who were with Captain Gass were attempting to feed more hose into the structure. There was a rapid buildup of heat and the hoseline seemed to drop. The firefighters exited the building and reported this situation to the Chief. Two Rapid Intervention Teams (RIT) were formed and, after four attempts, the second team was successful in recovering Captain Gass. Captain Gass was equipped with full structural protective clothing and a manually activated PASS device. The PASS was found in the "off" position. Captain Gass was located about 18 feet inside the front door of the structure. Captain Gass was removed from the structure approximately 20 minutes after his arrival on the scene. The cause of death was listed as smoke and soot inhalation with greater than 80 percent total thermal injury.

Additional information about this incident may be found in NIOSH Fire Fighter Fatality Investigation F2000-09.

Firefighter Fatalities in the United States in 2000

January 27, 2000
Robert Boy Ketelsen, Firefighter
Age 59, Volunteer
Westbrook Fire Department, Connecticut

Firefighter Ketelsen and members of his fire department responded to an automatic fire alarm. The alarm turned out to be unfounded. Fire department members returned to the fire station and placed the fire apparatus in-service. Less than 2 minutes after departing the fire station for home in his personal vehicle, Firefighter Ketelsen was struck with a heart attack. He managed to pull off the road into a parking lot before he became unconscious. Firefighter Ketelsen was found in full cardiac arrest when members of his fire department arrived. He was transported to a local hospital where he was pronounced dead. Firefighter Ketelsen had a history of heart problems.

February 6, 2000
Douglas George Stevens, Training Officer
Age 42, Volunteer
Story City LaFayette Township Volunteer Fire Department, Iowa

Training Officer Stevens was working on the scene of a residential structure fire. He climbed a ground ladder, used a halligan tool to remove a section of siding, and then continued on to the roof to prepare to perform additional ventilation. Training Officer Stevens descended the ladder and walked toward a backup hose team that was standing by outside of the residence. As he neared the other firefighters, Training Officer Stevens collapsed due to an apparent heart attack. Paramedics standing by on the scene initiated care immediately. Advanced Life Support (ALS) medical care was provided during a 17-minute transport to the hospital to no avail. Training Officer Stevens was pronounced dead shortly after arriving at the hospital. The cause of death was listed as occlusive coronary artery disease. The cause of the fire was an overheated wall next to a chimney.

Additional information about this incident may be found in NIOSH Fire Fighter Fatality Investigation F2000-14.

Firefighter Fatalities in the United States in 2000

February 11, 2000
Paul Eugene Cooper, Firefighter
Age 26, Volunteer
Hoopa Volunteer Fire Department, California

Firefighter Cooper was responding as the driver of an engine apparatus enroute to a motor vehicle accident on a narrow two-lane road. While the engine was about to negotiate a slight left curve, a car approached from the other direction straddling the line between the two lanes.

Firefighter Cooper moved the apparatus to the right side of the road to avoid a collision, and the engine's right tires left the pavement and drove onto a soft grassy shoulder. The truck continued on the shoulder, began to fishtail, glanced off of a power pole on the right side of the road, veered to the left out of control, and struck a large oak tree. Firefighter Cooper was trapped behind the steering wheel, and the firefighter who had been a passenger in the apparatus was ejected. Hoopa firefighters, assisted by other firefighters, extricated Firefighter Cooper after almost an hour of effort. Both firefighters were transported to the hospital by Hoopa Ambulance.

Although he was alert and conscious during the extrication, Firefighter Cooper entered a coma in the hospital. He never regained consciousness and died on February 14th, his 27th birthday. Neither Firefighter Cooper or his passenger were wearing a seat belt.

Additional information about this incident may be found in NIOSH Fire Fighter Fatality Investigation F2000-17.

February 13, 2000
Richard Owen Spink, Lieutenant
Age 48, Career
Fort Campbell Fire Department, Kentucky

Lieutenant Spink had just completed participating in a live burn structural training session. Lieutenant Spink was participating in a critique of the training when he was struck with a massive heart attack. There was a 4-minute delay in the response of an ambulance due to communications problems.

Lieutenant Spink had a number of health factors including prior heart attacks or chest pain, high blood pressure, diabetes, smoking, and weight problems.

Firefighter Fatalities in the United States in 2000

February 14, 2000
Lewis Evans Mayo, III, Firefighter
Age 44, Career
Houston Fire Department, Texas

Kimberly Ann Smith, Firefighter
Age 30, Career
Houston Fire Department, Texas

Firefighter Mayo and Firefighter Smith responded with Engine Company 76, three other engines, two ladder companies, two chief officers, an ambulance, and support staff to the report of a fire in a McDonald's restaurant. The fire was reported at 4:30 a.m. Engine Company 76 was comprised of a captain, a fire apparatus operator, and two firefighters. Engine 76 was the first fire fighting unit on the scene 8 minutes later and reported 6-foot flames visible from the roof. The flames appeared as if they might be venting from an exhaust fan, possibly indicating a grease fire.

The captain ordered his firefighters to advance an attack line into the interior of the structure for fire control. No fire was visible in the interior of the restaurant. The firefighters from Engine 76 were joined by other firefighters who also advanced attack lines to the interior. At 4:52 a.m., the incident commander ordered all firefighters out of the building in order to transition to a defensive attack mode. The flames visible from the roof had grown to 30 feet in height, and fire had become visible in the kitchen area of the restaurant.

Moments later, the captain from Engine 76 concluded that his firefighters were missing and notified the incident commander. A second alarm was requested at 5:02 a.m. and rescue attempts were begun. A number of rescue attempts were made.

At 5:27 a.m., the incident commander struck a third alarm. Shortly thereafter, a ladder company opened the rear door of the restaurant and made access to the back of the kitchen area. A PASS device had been heard alarming in the kitchen area, and a firefighter was able to see a downed firefighter as he looked into the back door. Firefighter Mayo was discovered with his facepiece in-place, his regulator not connected to the facepiece, and with his SCBA partially removed and entangled in wires. He was removed, treated at the scene, in the ambulance, and at the hospital. Despite these efforts, he was pronounced dead at the hospital.

Given the amount of time that had passed and the likelihood that Firefighter Smith was buried in debris, the search effort transitioned into a recovery mode. Firefighter Smith was found at approximately 7:13 a.m. within 6 feet of the rear door of the restaurant. She was entangled in wires and a pair of wire cutters were found near her body. She was wearing an SCBA but the status of her facepiece and regulator could not be determined.

Firefighter Fatalities in the United States in 2000

Both firefighters died of asphyxia due to smoke inhalation. Firefighter Mayo's carboxyhemoglobin level was found to be 26 percent and the level for Firefighter Smith was found to be 52 percent.

The fire was intentionally set by a group of juveniles attempting to conceal a burglary attempt. Four individuals were convicted of crimes with sentences ranging from 2 to 35 years.

Additional information about this incident may be found in NIOSH Fire Fighter Fatality Investigation F2000-13.

February 19, 2000
James D. Geiger, Firefighter-EMT
Age 55, Volunteer   Part Paid
City of Defiance Fire & Rescue, Ohio

Firefighter-EMT Geiger responded with other members of his fire department to a sledding accident. Firefighter-EMT Geiger assisted with patient packaging and helped carry the patient through deep snow to an ambulance. Firefighter-EMT Geiger left the scene at the conclusion of the incident in his private vehicle. Firefighter-EMT Geiger suffered a heart attack as he arrived at his home; his car struck a propane tank. Firefighters were called to the scene and transported Firefighter-EMT Geiger to the hospital where he died shortly after arrival. The autopsy cited severe occlusive coronary artery disease as the cause of death.

February 21, 2000
Evangelino Soto Rodriguez, Sergeant
Age 53, Career
Puerto Rico Fire Department, Puerto Rico

Sergeant Rodriguez was called to the scene of an arson-caused lumberyard fire. Sergeant Rodriguez had just finished attaching a hoseline to a hydrant and began to cross the road back to his engine company. As he crossed the road, a car operated by a drunk driver struck him. The cause of death was listed as severe multiple trauma.

Firefighter Fatalities in the United States in 2000

February 29, 2000
Robert Jeffery Jackson, Firefighter
Age 35, Volunteer
Harmony Volunteer Fire Department, Atoka, Oklahoma

Firefighter Jackson was responding to a mutual aid grass fire in a neighboring community. Firefighter Jackson was the sole occupant and driver of a three-quarter-ton four-by-four brush truck. As Firefighter Jackson responded, he encountered a sedan traveling in the opposite direction. As the sedan crested a hill, the driver lost control, skidded approximately 258 feet, crossed the center of the road, and struck the brush truck head on. Firefighter Jackson attempted to avoid the collision by pulling to the side of the road. The emergency lights on the brush truck were activated. Firefighter Jackson was wearing a seat belt. Firefighter Jackson's speed was estimated at 30-35 miles per hour and the speed of the sedan was estimated at 70 miles per hour.

After the collision, both vehicles caught fire. The fire was reported to the Harmony Volunteer Fire Department, and they responded to the incident. Upon arrival, both vehicles were found to be fully involved in fire. The cause of death was listed as massive blunt chest trauma with burns noted as another significant medical condition.

Firefighter Fatalities in the United States in 2000

March 4, 2000
David Paul Sutton, Firefighter
Age 27, Volunteer
Fraser Department of Public Safety, Michigan

Firefighter Sutton responded, along with other members of his public safety department, to a working apartment fire. While they were engaged in suppression of the first fire, another apartment fire was reported in a building across the street from the original fire. Since no fire apparatus was available to respond, Firefighter Sutton and other firefighters responded in a van to the scene. Police officers were in the process of evacuating the building. A resident in need of rescue had been spotted at a second story window. Mutual aid fire companies were responding but not yet on the scene. The smoke condition at the entrance to the apartment building was light, with heavier smoke and heat on the second floor. Fire at the top of the stairs was observed by one firefighter. Firefighter Sutton and another firefighter, equipped with full-protective clothing and SCBA, entered the building to effect the rescue. Witnesses outside the building reported that the resident disappeared from the window as if she had been reached by firefighters. Within seconds, a flashover occurred, trapping the resident and the two firefighters. Both firefighters managed to reach a bathroom at the rear of the apartment, but they were unable to get through the window with their SCBA in-place. Firefighter Sutton was observed by other firefighters at the window, and a rescue effort was mounted.

Two firefighters shed their SCBA and entered the bathroom from ground ladders. Firefighter Sutton was removed after his SCBA was cut from him. The low pressure hose on his SCBA had burned through. The other firefighter was located in the bathtub and removed. Both firefighters were transported to the hospital. Firefighter Sutton was pronounced dead at the hospital. The cause of death was listed as asphyxiation. The injured firefighter sustained major burns and was hospitalized for 6 months. The resident of the apartment also died.

The fire was caused when an arsonist(s) ignited combustibles on the first and second floors of the apartment building. This fire was one of six arson fires that occurred in the same general area over 2 days.

Additional information about this incident may be found in NIOSH Fire Fighter Fatality Investigation Report F2000-16.

Firefighter Fatalities in the United States in 2000

March 6, 2000
Robert W. Buhler, Firefighter
Age 62, Volunteer
Delmont Volunteer Fire Department, South Dakota

Firefighter Buhler and members of his fire department were fighting a wildland fire. The fire was the result of a controlled field burn that was being conducted by some local citizens that got out of control. The conditions were dry with winds of 40 miles per hour. The fire was in a very deep winding ravine. Hose was being added to an attack line when a wind gust blew up an area that had been thought to be previously extinguished. The fire spread rapidly up a hill and engulfed Firefighter Buhler. Firefighter Buhler had responded directly to the scene from a nearby town and was not wearing protective clothing. Firefighter Buhler was severely burned over 60 to 80 percent of his body and died on March 16, 2000. Another firefighter, who was near Firefighter Buhler at the time of the blowup and who was equipped with protective clothing, received minor injuries.

Additional information about this incident may be found in NIOSH Fire Fighter Fatality Investigation F2000-22.

March 6, 2000
Donald R. Wilson, Assistant Chief
Age 50, Volunteer
Herrick Fire Protection District, Illinois

Assistant Chief Wilson was on the roof of a residence that was involved in fire. The fire started when a garbage fire extended through brush to a house. He had just chopped a hole in the roof to determine if the fire had spread to the attic. He was seen lying on the roof. Firefighters found that Assistant Chief Wilson had suffered a heart attack. Despite their efforts, Assistant Chief Wilson was pronounced dead at the scene.

March 7, 2000
Jerry Wayne Coppin, Training Officer
Age 56, Volunteer
Okay Volunteer Fire Department, Oklahoma

Training Officer Coppin and members of his department responded to assist with storm watch duties in their community. Toward the end of the storm, Training Officer Coppin suffered a cerebral hemorrhage as he sat in his pickup truck and operated a radio. Medical aid was provided immediately by first responders and ambulance personnel. Training Officer Coppin was transported to a hospital but died on March 11, 2000.

Firefighter Fatalities in the United States in 2000

March 8, 2000
William M. Blakemore, Private
Age 48, Career
Memphis Fire Department, Tennessee

Javier Lerma, Lieutenant
Age 41, Career
Memphis Fire Department, Tennessee

Engine 55, a four-person engine company including Lieutenant Lerma and Private Blakemore, responded to the report of a residential structure fire along with other units. Engine 55 was the first unit on the scene and reported a working house fire. Lieutenant Lerma stepped from the apparatus to perform a size up of the fire and was immediately shot by a gunman who had been hiding in the garage of the house. The gunman continued to fire, striking Private Blakemore as he sat in the back of the pumper preparing his protective equipment. The driver of Engine 55 moved the apparatus forward out of the danger zone. The fourth member of the crew had been on the opposite side of the apparatus preparing his protective clothing and jumped onto the running board as the apparatus was moved to safety. As the gunman engaged and killed a deputy sheriff, firefighters moved Lieutenant Lerma to a safe area and began treatment.

Lieutenant Lerma was pronounced dead at the scene and Private Blakemore died in an ambulance enroute to the hospital. Both were killed by shotgun blasts to the head.

After the scene was secured, firefighters extinguished the fire in the house and discovered the body of the gunman's wife. A deputy sheriff was also killed.

Lieutenant Lerma's father was killed in the line-of-duty in 1977. Lieutenant Lerma was carried to his rest in a fire truck named after his father.

The gunman was an off-duty Memphis firefighter. He was responsible for the fire in the house, and he was likely the person who reported the fire to the Memphis Fire Department through a 9-1-1 call.

Firefighter Fatalities in the United States in 2000

March 13, 2000
Jessie Lamar Y'Barbo, Forestry Technician III
Age 54, Wildland Career
Texas Forest Service, Texas

Forestry Technician Y'Barbo was participating in the controlled (prescribed) burn of a 35-acre block of mature pine forest, with low understory vegetation. Forestry Technician Y'Barbo was operating a 1985 Honda 250 All-Terrain Vehicle (ATV), which had been procured by the Forest Service as Federal government property. The ATV had a shop-made holder at the rear that accommodated a drip torch. Forestry Technician Y'Barbo was wearing brush gear.

As Forestry Technician Y'Barbo ascended an incline while riding the ATV, the vehicle overturned backward. As Forestry Technician Y'Barbo struggled to free himself, he accidentally kicked off the cap on the ATV's fuel tank. Fuel splashed on Forestry Technician Y'Barbo and was quickly ignited by the drip torch.

Other firefighters quickly came to Forestry Technician Y'Barbo's aid, and he was transported to a hospital to undergo burn treatment. Despite the efforts of his fellow firefighters and hospital personnel, Forestry Technician Y'Barbo died on April 7, 2000. The cause of death was listed as complications of thermal injuries.

An Accident Review and Mitigation Report prepared by the Forest Service recommended that all Forest Service ATVs be equipped with threaded caps (the cap on Forestry Technician Y'Barbo's ATV was only equipped with a quarter-turn cap); fuel tank venting and overflow tubes; operator training, equipping supervisors with first aid kits that include fire blankets; the consideration of replacement of older, narrow track ATVs; and the installation of rollover protection on ATVs.

Forestry Technician Y'Barbo was the past Fire Chief of the Kirbyville Fire Department.

March 15, 2000
Mike Shortt, Fire Chief
Age 44, Career
Weaverville Volunteer Fire Department, California

Chief Shortt was acting as an instructor for an evening drill on ground ladder placement and raises. Chief Shortt acted mainly as an instructor/coach and did not personally participate in much actual ladder handling. During the class, Chief Shortt collapsed, the victim of a heart attack. Medical aid was provided immediately by firefighters attending the class. Chief Shortt was transported to the hospital and survived until March 31, 2000.

Firefighter Fatalities in the United States in 2000

March 17, 2000
David Clements Sharp, II, Firefighter/Engineer
Age 31, Career
Fayetteville Fire/Emergency Management Department, North Carolina

Firefighter/Engineer Sharp responded to an automatic fire alarm as the driver and lone occupant of a 1993 Pierce Arrow 100-foot ladder tower truck. The first unit on the scene found a system malfunction and canceled all other responding fire apparatus, including the ladder truck operated by Firefighter/Engineer Clements.

As he returned to the fire station, Firefighter Clements came upon a railroad crossing that was blocked by traffic control devices as a slow-moving freight train passed. As the last car of the train passed, it stopped just past the intersection. Firefighter Clements drove the ladder truck around the traffic control arm and attempted to cross the tracks. As he passed the freight train, a passenger train headed in the opposite direction of the freight train struck the left front of the fire truck. After the collision, the truck spun around and Firefighter/Engineer Clements was thrown from the vehicle. Firefighter/Engineer Clements came to rest under the rear tires of the ladder truck. He was pronounced dead at the scene. The cause of death was listed as multiple blunt trauma. Firefighter/Engineer Sharp was not wearing a seat belt.

Witness statements indicated that Firefighter/Engineer Sharp's view of the oncoming passenger train was likely blocked by the freight train. The train involved in the collision was moving at 30 miles per hour just prior to the impact.

Additional information about this incident may be found in NIOSH Fire Fighter Fatality Investigation F2000-19.

March 18, 2000
Frederick L. Brain, Fire Police Officer
Age 76, Volunteer
Miller Place Fire Department, New York

Fire Police Officer Brain was directing traffic at the scene of a motor vehicle collision. After approximately an hour on the scene, Fire Police Officer Brain collapsed. The fire chief reached Fire Police Officer Brain and found him gasping for breath but had a weak pulse. Firefighters on the scene applied an Automatic External Defibrillator (AED) and the unit delivered a shock. A firefighter performed CPR on Fire Police Officer Brain as an ambulance was summoned.

Fire Police Officer Brain was transported to the hospital with a pulse. He lapsed into a coma and died on April 24, 2000.

Firefighter Fatalities in the United States in 2000

March 27, 2000
Kevin Francis Sterenchuk, Administrative District Chief
Age 48, Career
Cedar Rapids Fire Department, Iowa

District Chief Sterenchuk was in his office performing administrative duties. He had not been feeling well an hour prior and had asked the department's EMS coordinator to take his blood pressure. His blood pressure was high, and he was also complaining of an ache in his right elbow similar to carpal tunnel pain. He agreed to have his blood pressure checked later.

The department's EMS coordinator, a captain, later walked past District Chief Sterenchuk's office and noticed that he was having difficulty breathing. The captain directed a staff member to call 9-1-1 and retrieved medical equipment from a reserve fire apparatus that was in the same building.

Despite immediate advanced life support medical aid by firefighters and ambulance personnel, District Chief Sterenchuk succumbed to a heart attack.

March 28, 2000
Michael "Mike" Russell Queen, Fire Chief
Age 30, Volunteer
Clayton Fire Department, Georgia

Fire Chief Queen was assisting with hose testing at his fire station. During the testing process, a 2½ -inch hose separated from its coupling. High-pressure water struck Chief Queen and propelled him into a fire truck that was parked nearby. Chief Queen suffered a fatal blow to the head as he hit the apparatus. A Firefighter/EMT began treatment immediately and Chief Queen was transported to the county hospital. He was pronounced dead approximately 45 minutes after the accident.

Firefighter Fatalities in the United States in 2000

March 31, 2000
Kendall O. Bryant, Firefighter/EMT
Age 36, Paid-on-Call
Layton City Fire Department, Utah

Firefighter/EMT Bryant and members of his department were dispatched to the report of a residential fire. Upon arrival, Firefighter/EMT Bryant's captain reported a working fire with flames and smoke visible from the garage. The captain ordered his firefighters to extinguish the fire in the garage, and the fire was knocked down within 5 minutes of their arrival on-scene.

The captain then instructed Firefighter/EMT Bryant and another firefighter to enter the structure with a hoseline to search for victims, fire extension, and to begin to ventilate the structure. The firefighters were met with dark smoke but no visible flame when they entered the structure. They began a left hand search and proceeded to the second floor of the structure. The second floor contained bedrooms and was directly above the garage. A lieutenant joined the firefighters on the second floor by following the hoseline. As the firefighters searched the bedrooms, there was a rapid buildup of heat. A red glow was visible at the bottom of the stairs, cutting off the team's escape route. The decision was made to follow the hoseline back out of the structure since the firefighters were unsure about the presence of windows in the bedrooms and the stairway was small. Firefighter/EMT Bryant was the last in line as the firefighters made their way to safety.

As the firefighters emerged from the house, the lieutenant removed his facepiece and told other firefighters that Firefighter/EMT Bryant was supposed to be right behind him but had not exited the structure with him. The incident commander ordered an accountability report and it was discovered that Firefighter/EMT Bryant was missing. A second crew of firefighters entered the residence through the front door but could not climb the stairs because they appeared to be collapsed and were heavily involved in fire. The incident commander ordered a ladder raised to provide firefighters with access to a roof area which led to the bedroom windows. Two firefighters entered the second floor of the structure and searched two bedrooms. A sound believed to be Firefighter/EMT Bryant's PASS device was located but turned out to be a smoke alarm.

The firefighters saw a light in the bedroom across the hall and found that it was a flashlight that was carried by Firefighter/EMT Bryant. Firefighter/EMT Bryant was found on his knees on the floor with his facepiece removed. His SCBA cylinder was found to be empty and his protective hood was found over his mouth and nose, most likely in an attempt to filter air to breathe. His PASS device was found in the "off" position. Firefighter/EMT Bryant was removed by firefighters through a window and lowered to the ground into the care of waiting paramedics.

Firefighter/EMT Bryant received aggressive resuscitation efforts at the scene, in the ambulance, and in a hospital emergency room, to no avail. Firefighter/EMT Bryant was pronounced dead in the emergency room. The cause of death was later listed as smoke and soot inhalation and acute carbon monoxide poisoning. Firefighter/EMT Bryant's blood carboxyhemoglobin level was found to be 25 percent at the time of his death. This level does not actually reflect the level that had been present in his blood since the level was reduced by resuscitation efforts.

Firefighter/EMT Bryant was a career firefighter in Ogden, Utah. Two other firefighters were injured. The fire was caused by a droplight that had been hung near a cardboard box that was being used as part of a dog's bed.

Additional information about this incident may be found in NIOSH Fire Fighter Fatality Investigation F2000-23.

April 4, 2000
David Anthony Maisano, Captain
Age 38, Volunteer
Tritown Fire Department, Mio, Michigan

Captain Maisano and members of his fire department responded to a report of smoke in a residential structure. After removing the contents of the fireplace and checking the attic and roof of the structure, it was determined that smoke was escaping the chimney in the attic and causing the smoke buildup. Firefighters were stowing equipment in preparation for their return to the station. Captain Maisano was attaching an elastic cord to secure a ground ladder to the truck when the cord snapped and struck him in the face. Captain Maisano fell to the ground from a height of approximately 9 feet, fracturing a wrist and causing severe back pain.

Captain Maisano was transported to the hospital by ambulance. He was treated there for 3 days and was discharged. He was unable to sleep in bed due to pain, so he slept in an easy chair located in his home on the night that he was released from the hospital, April 7th. He was last known to be alive at approximately 2:30 a.m. on April 8th, but was discovered dead by his wife in the morning.

The autopsy attributed his death to pneumonia. The autopsy also noted that the use of multiple prescribed pain relief medications may have initiated an element of respiratory depression.

Firefighter Fatalities in the United States in 2000

April 7, 2000
James Ted Griffith, Firefighter/Training Officer
Age 25, Volunteer
Winterset Volunteer Fire Department, Iowa

Firefighter/Training Officer Griffith and members of his department were called to the scene of a grass fire. The fire started when salvage workers ignited grass and nearby wood as they worked to dismantle two old, rusting 12,000-gallon elevated fuel storage tanks. The grass fire was extinguished, and the salvage workers decided to use a blowtorch to cut a small hole near the drain of the tank that had already been pulled to the ground. The hole was intended to allow the attachment of a tow chain, which would be used to pull the tank to a salvage yard. As the hole was being made with a blowtorch, the tank emitted a hissing sound and suddenly exploded.

Firefighter/Training Officer Griffith was killed instantly when he was struck by flying debris. The cause of death was listed as multiple blunt trauma.

The top of the tank, which was torn away in the explosion and weighed over 900 pounds, flew over 114 feet before coming to rest. A salvage worker was also killed, eight firefighters and a civilian received injuries. Analysis of the tank contents revealed that the tank contained residual gasoline and other petroleum products.

Additional information about this incident may be found in NIOSH Fire Fighter Fatality Investigation F2000-25.

Firefighter Fatalities in the United States in 2000

April 11, 2000
Michael R. Baughn, Firefighter
Age 46, Paid-on-Call
Washington Fire/Rescue, Ohio

Firefighter Baughn and 30 other members of his department were participating in search and rescue training in the basement of an office building. Teams of two firefighters, equipped with structural protective clothing and SCBA, were doing crawl through searches in areas that were obscured with nontoxic smoke. After completing an exercise that lasted approximately 20 minutes, Firefighter Baughn sat in a basement hallway, removed his SCBA facepiece, rested, and waited for other firefighters to complete the exercise.

When the other firefighters emerged from the exercise, Firefighter Baughn began to climb the stairs from the basement to the ground level of the building. As he reached the first landing in the stairwell, Firefighter Baughn suddenly collapsed. Firefighters found Firefighter Baughn pulseless and gasping for air. CPR was started immediately and an ambulance was called to the scene.

Arriving ambulance personnel from the local ambulance squad began ALS treatment but were unable to deliver a shock to Firefighter Baughn through their defibrillator. Firefighter Baughn was transported immediately to the closest hospital, a 2-minute ambulance ride away from the training site. Resuscitation efforts were continued in the emergency room for 30 minutes until Firefighter Baughn was pronounced dead.

The cause of death was listed as Cardiomegaly (enlarged heart) acute cardiac arrhythmia. Firefighter Baughn had a history of high blood pressure. He took two medications for this disease, but did not take them regularly.

The energy selector for the defibrillator was found to be in the "0" mode. The defibrillator will not deliver a shock unless an energy level greater than zero for the shock is selected. Additional defibrillator training was recommended for squad members.

Additional information about this incident may be found in NIOSH Fire Fighter Fatality Investigation F2000-24.

Firefighter Fatalities in the United States in 2000

April 20, 2000
Rickey Levi Davis, Firefighter/Paramedic
Age 33, Career
Center Point Fire/Rescue, Alabama

Firefighter/Paramedic Davis and members of his department were dispatched to a report of a fire in a single-family residential structure that included a full basement. Upon arrival, firefighters found heavy smoke showing from the structure and found that the fire was in the basement. Firefighters attempted to reach the fire through the garage door (which opened into the basement) but were unsuccessful in locating the seat of the fire. A positive-pressure fan was placed at the garage door. Another team of three firefighters, including Firefighter/Paramedic Davis, advanced an attack line through the front door of the residence. On their initial entry into the residence, they were unable to locate any fire. The crew withdrew, found that a positive-pressure fan had been placed at the front door, and returned to explore another area of the house.

Firefighter/Paramedic Davis was at the nozzle as the hoseline was advanced into the second entry on the main floor of the residence. As the line was advanced, Firefighter/Paramedic Davis fell through the floor into the area of the basement that was involved in fire. Other firefighters helped Firefighter/Paramedic Davis as he attempted to jump back to the first floor from the basement but his efforts were unsuccessful. Firefighters attempted to lower a scuttle hole ladder into the hole but the location of the hole and the sagging of the first floor into the basement prevented its use. Firefighters instructed Firefighter/Paramedic Davis to use the hoseline to protect himself as they attempted to rescue him through the basement.

An attack team entered the basement and fought their way to the room that contained Firefighter/Paramedic Davis. He was removed from the basement and received ALS medical treatment immediately. He was transported by ground and air ambulances to a hospital in nearby Birmingham. He was treated in the emergency room but was pronounced dead.

The cause of death was listed as hyperthermia (thermal injuries). The carboxyhemoglobin level that was found in Firefighter/Paramedic Davis' blood was less than 5 percent. He was burned over roughly one-third of his body.

It is estimated that 12 to 15 minutes passed from the time Firefighter/Paramedic Davis fell into the basement until he was located and removed from the structure. Firefighter/Paramedic Davis was the first firefighter fatality for Center Point Fire/Rescue.

Firefighter Fatalities in the United States in 2000

April 26, 2000
Robert Cowey Brannon, Jr., Lieutenant
Age 43, Career
Bluefield Fire Department, West Virginia

Lieutenant Brannon responded with his engine company to a report of a structure fire in a residence. Lieutenant Brannon was part of a crew of three that responded on his company. Arriving companies found a working fire on the first and second floors of a 1½ -story wood frame house. Lieutenant Brannon took command of the incident and ordered an attack line through the front door. After completing a 360-degree walk around the involved structure, Lieutenant Brannon assisted with the deployment and advancement of a second attack line. The line was stretched to the first floor of the house and was used to control hot spots. Lieutenant Brannon noticed an air leak on his SCBA. The leak was controlled by another firefighter inside the house, but Lieutenant Brannon found that he was out of air and needed a new cylinder.

Lieutenant Brannon exited the house, spoke momentarily with the fire chief, and proceeded to his apparatus to get his cylinder changed. He kneeled at the truck to allow the driver/operator to replace his cylinder. The driver/operator asked Lieutenant Brannon if he was okay, Lieutenant Brannon responded that he needed a new cylinder, turned his head, and collapsed.

Other firefighters and on-scene paramedics immediately came to Lieutenant Brannon's aid. Lieutenant Brannon's personal protective equipment was removed and ALS care was initiated. He was defibrillated several times on the scene and was transported to a local hospital by the Bluefield Rescue Squad. Lieutenant Brannon was revived in the emergency room but died on May 4, 2000. The cause of death was listed as artherosclerotic coronary artery disease.

Additional information about this incident may be found in NIOSH Fire Fighter Fatality Investigation F2000-34.

Firefighter Fatalities in the United States in 2000

April 29, 2000
David John Liston, Smokejumper
Age 28, Wildland Part-Time
Bureau of Land Management - Alaska Fire Service, Alaska

Smokejumper Liston was participating in mandatory annual recertification practice parachute jumps in preparation for the upcoming wildland season. He had completed three jumps. During the fourth jump of the day, Smokejumper Liston's parachute failed, and he plunged 3,000 feet to his death. Emergency medical care was provided immediately by other smokejumpers trained as emergency medical technicians to no avail. Smokejumper Liston's cause of death was listed as multiple impact (deceleration) injuries. All smokejumping activities in Alaska and Idaho were halted for over 2 months as this incident was investigated.

The investigation revealed that the parachute malfunction was characterized as a "drogue in tow" meaning that the drogue chute deployed but did not release on demand in order to deploy Smokejumper Liston's main parachute. Smokejumper Liston then followed emergency procedures and manually deployed his reserve parachute. During this action, the reserve pilot chute became entangled with the drogue bridle (the line which attaches the drogue to the main parachute) thereby preventing both the main and reserve canopies from deploying.

Smokejumper Liston was the first parachute-related fatality for the Bureau of Land Management in 40 years.

April 29, 2000
L.C. Merrell, Lieutenant
Age 43, Career
Chicago Fire Department, Illinois

Lieutenant Merrell was in command of a truck company responding with lights and siren to a still alarm. Lieutenant Merrell was riding in the right front seat of the apparatus, and he was not wearing a seat belt.

The truck entered a four-way stop intersection and was broadsided by a pickup truck that ran the stop sign. Lieutenant Merrell was thrown from the cab and struck the pavement. Immediate medical care was provided by other firefighters and medical personnel, but Lieutenant Merrell was pronounced dead at the scene. The cause of death was listed as blunt head trauma. The truck company slowed prior to entering the intersection.

Four other firefighters and nine civilians in two vehicles were injured. The still alarm turned out to be a false alarm. The Chicago Fire Department Commissioner was quoted as saying that Lieutenant Merrell could have survived the accident if he had been wearing a seat belt. The driver of the pickup was ticketed for speeding and failure to stop.

Firefighter Fatalities in the United States in 2000

April 30, 2000
Arnold Blankenship, III, Second Assistant Chief
Age 27, Volunteer
Greenwood Volunteer Fire Company #1, Inc., Delaware

Assistant Chief Blankenship and other members of his department were participating in a training/ demolition burn of a two-and one-half story wood frame dwelling.

According to the fire chief, the plan for the day was to do small, one room burns to evaluate a saw, and then to completely burn the house. After a series of small fires were extinguished on the first floor of the house, preparations were made for the demolition burn.

The plan for the final fire was to ignite the attic, then ignite the first floor, evacuate the house, and allow it to burn completely. Water curtain nozzles were set up on the exterior of the house to protect trees that were in the proximity of the house. Assistant Chief Blankenship went into the attic of the house and used a small garden-type sprayer to distribute diesel fuel in the attic. As fires were ignited inside an attic room, Assistant Chief Blankenship used the sprayer to "accelerate" the fire. With the exception of Assistant Chief Blankenship, all firefighters had left the attic space and were proceeding to the first floor of the structure. A firefighter waiting at the base of the attic stairs for Assistant Chief Blankenship noted fire and smoke coming from the attic. When firefighters reached the exterior of the structure, they notified the fire chief that Assistant Chief Blankenship was missing and possibly trapped.

As some firefighters attempted to suppress the fire, other firefighters used a ground ladder to access the second floor of the house in an attempt to rescue Assistant Chief Blankenship. After several attempts, firefighters followed the sound of an activated PASS device and were able to reach Assistant Chief Blankenship. They were unable to remove him as portions of the collapsed roof covered him. Mutual aid firefighters arrived and were able to locate and remove Assistant Chief Blankenship about an hour after the time he was reported missing. He was obviously deceased. The cause of death was later listed as asphyxiation and burns.

A Major Incident Response Team from the State Fire Marshal's Office conducted a thorough review of the incident. The review concluded that there were several contributing factors to Assistant Chief Blankenship's death. The factors listed were:

Assistant Chief Blankenship remained in the attic too long, despite the urging of at least two other firefighters to exit the attic due to deteriorating fire conditions.

The initiation of several fires simultaneously in the west playroom and the south wing by orders of Assistant Chief Blankenship.

The confined space of the attic construction caused the unstable conditions in the attic resulting in raising the ambient temperature; thereby causing volatile conditions in the room where the firefighters were present.

The use of atomized diesel fuel through a garden sprayer by Assistant Chief Blankenship directly on a free burning fire in the south wing of the attic resulted in the flash fire that enveloped the attic, and ultimately claimed his life.

May 7, 2000
Carl Ray Payne, Pilot
Age 66, Wildland Part-Time
Payne Flying Service, under contract to the United States Department of the Interior for the Texas Forest Service, Lufkin, Texas

Pilot Payne had just dropped a load of fire retardant on a fire outside of Fort Stockton, Texas. The aircraft was an Air Tractor AT-802A. After the drop, Pilot Payne was circling two radio antenna (280 feet and 310 feet) when the outboard 5 feet of his right wing struck the antenna guide cables and support structure. Pilot Payne was able to level the aircraft and continued to fly; however, the aircraft was compromised and struck trees and terrain before coming to rest. Cable strands from the antenna towers were found among the aircraft wreckage. The cause of death for Pilot Payne was listed as multiple severe trauma.

In a letter written to a friend just before his death, Pilot Payne said that the year 2000 was going to be his last year flying for a living and that he was going to relax with his hobbies and his family. He said that he was told by a good friend that life was not a dress rehearsal, it was the real thing, and that he was going for it.

Additional information related to this incident may be found in National Transportation Safety Board Accident Investigation FTW00TA142.

May 8, 2000
Kenneth Jesse, Fire Police Officer
Age 80, Volunteer
Harford Volunteer Fire Company, Pennsylvania

Fire Police Officer Jesse had responded with members of his department to a vehicle fire on the interstate. On the scene of the vehicle fire, he blocked traffic to protect firefighters engaged in control of the fire. When the fire was extinguished, Fire Police Officer Jesse returned home. Upon his arrival at home, Fire Police Officer Jesse told his wife that he felt dizzy and had a lump in his stomach. His wife went inside their house for a moment to prepare to take him to the doctor and found that he had collapsed in their carport when she returned outside. His wife called 9-1-1, and firefighters from Fire Police Officer Jesse's department responded and began CPR. Despite their efforts, Fire Police Officer Jesse died of a heart attack.

Firefighter Fatalities in the United States in 2000

May 13, 2000
Arthur E. Nilson, Jr., Company Member
Age 75, Volunteer
Walker Township Volunteer Fire Department, Pennsylvania

Company Member Nilson suffered a fatal heart attack as he helped to set up for a company banquet.

May 15, 2000
Leo Koponen, Air Attack Pilot
Age 49, Wildland Contract Part-Time
Courtney Aviation, Columbia, California   Under Contract to the United States Forest Service

Samuel James Tobias, Air Tactical Group Supervisor
Age 47, Wildland Career
United States Forest Service, Lincoln National Forest, Alamagordo, New Mexico

Air Attack Pilot Koponen and Air Tactical Group Supervisor Tobias were beginning a reconnaissance flight to look for any fires that may have spread from the on-going Scott-Able fire. The aircraft was a Cessna T337C. Approximately 6 minutes after takeoff, black smoke was noted in the vicinity and calls were received from local residents reporting smoke in Alamo Canyon. A helicopter flying in the area confirmed that there was a downed aircraft and that there appeared to be no survivors. Investigators found that the aircraft had crashed nose first and that the deaths of both firefighters were immediate.

The weather at the time of the flight was clear with a light wind. Additional information related to this incident may be found in National Transportation Safety Board Accident Investigation DEN00GA089.

Firefighter Fatalities in the United States in 2000

May 27, 2000
Evan N. Shirk, Firefighter
Age 27, Volunteer
Moreau Fire Protection District, Missouri

Firefighter Shirk was the sole occupant and driver of an engine apparatus returning from a motor vehicle accident. The pumper was equipped with a 1,000-gallon water tank that was filled to 900 gallons. The accident turned out to be an unfounded report. As Firefighter Shirk returned to the station, the right front wheel drifted off of the road onto soft ground. According to the police accident report, Firefighter Shirk overcorrected and the right wheel struck a drainage culvert, causing the pumper to veer across the road, roll over several times, and catch fire. Firefighter Shirk was not wearing a seat belt and was ejected.

Firefighter Shirk was pronounced dead at the scene. The cause of death was listed as massive head trauma. Firefighter Shirk was the first line-of-duty death for his department. The police accident report listed driver inattention as a factor in the crash.

Additional information about this incident may be found in NIOSH Fire Fighter Fatality Investigation F2000-33.

May 31, 2000
Lyndell J. Smith, Firefighter
Age 46, Volunteer
Caldwell Fire Department, Kansas

Firefighter Smith was a passenger in a 1984 Jeep CJ7 command vehicle. The truck was equipped with extrication equipment and was responding with lights and siren in operation to a vehicle rollover with reports of serious injuries. Firefighter Smith was riding the hump between the driver and the front seat passenger. None of the truck's occupants were wearing seat belts. As the command vehicle overtook and passed a passenger car on the left side, the car turned into the command vehicle, striking it at the right rear tire. The command vehicle skidded across traffic, entered a ditch, overturned in a wheat field, and caught fire. All three occupants of the command vehicle were ejected. Firefighter Smith received fatal injuries, and the other occupants of the command vehicle received serious injuries. No autopsy was performed but the cause of death for Firefighter Smith was listed as closed head and chest trauma with exsanguination from mortal axillary wounds (blood loss).

No extrication was required at the scene of the original rollover call.

Firefighter Fatalities in the United States in 2000

June 4, 2000
George A. "Bo" Burton, Firefighter/Rotor Craft Pilot
Age 48, Wildland Career
Florida Division of Forestry

Firefighter/Rotor Craft Pilot Burton was fighting a fire near Fort Myers, Florida. He had been on the scene for approximately 1½ hours, performing reconnaissance, making water drops, and filling his external bucket from a local lake. Witnesses observed the helicopter in level flight headed back to the lake after a water drop. The helicopter was reported to suddenly bank deeply with its nose down. After a few seconds, the helicopter crashed in a cow pasture.

The cause of the crash has not been determined. The aircraft was a 1966 Bell UH1 (205).

Additional information related to this incident may be found in National Transportation Safety Board Accident Investigation MIA00GA184.

June 25, 2000
Whitney C. Teehan, Jr., Captain
Age 66, Volunteer
Eastern Point Volunteer Fire Company #2, Groton, Connecticut

Captain Teehan and members of his department responded to a manufacturing facility for a report of fire. Captain Teehan was acting as the department's accountability officer on the scene of the incident. It was determined that a large dust cloud caused by a high-pressure air leak had been mistaken for smoke.

As companies prepared to return to quarters, Captain Teehan suffered a massive heart attack while seated in a pumper. Other firefighters began CPR and an Automatic External Defibrillator (AED) was used. The AED was unable to restore a rhythm, and Captain Teehan was transported to the hospital by ambulance. He was later pronounced dead.

Captain Teehan had suffered a previous heart attack on the scene of a car fire in 1997. An AED operated by a private fire department saved him that day and he went on to support their purchase and use of AED's by his own fire department in the future. Captain Teehan suffered a heart attack on the grounds of the Electric Boat plant, where he had been employed for 41 years prior to his retirement in 1993.

Firefighter Fatalities in the United States in 2000

July 2, 2000
Nathan Andrew Pescatore, Firefighter
Age 17, Volunteer
Lloydsville Volunteer Fire Department and Relief Association, Pennsylvania

Firefighter Pescatore was responding as the sole occupant and driver of his personal vehicle to a report of a structure fire. He crossed the centerline of the road as he entered a curve in the road. As he rounded the curve, he came upon a farm tractor approaching from the opposite direction. Firefighter Pescatore's view of the tractor as he drove into the curve was blocked by vegetation.

Firefighter Pescatore was unable to get back into his lane and struck the farm tractor head on. The loader bucket on the front of the tractor was driven through both driver's side roof posts and severely injured Firefighter Pescatore. Firefighters responding on mutual aid to the structure fire were diverted to the collision and were joined by Lloydsville firefighters at the scene. After Firefighter Pescatore was extricated, he was flown by helicopter to the hospital.

Firefighter Pescatore was pronounced dead at the hospital due to blunt force trauma to the head and chest.

July 15, 2000
Phillip Ridings, Firefighter
Age 52, Volunteer
Hornersville Volunteer Fire Department, Missouri

Firefighter Ridings was advancing a 1½-inch hoseline at the scene of a two-story residential fire. The home was originally built in 1917 and had been renovated several times. The cause of the fire was electrical.

As he worked, Firefighter Ridings became fatigued and suffered a fatal heart attack. Firefighter Ridings was a highly decorated Vietnam veteran.

Firefighter Fatalities in the United States in 2000

July 18, 2000
Steven Max Wilmot, Captain
Age 47, Career
Springfield Fire Department, Illinois

Captain Wilmot was a fire investigator working on the scene of a previous structure fire. As Captain Wilmot and another fire investigator worked on the scene, Captain Wilmot caught his foot on an object and fell forward onto his chest, landing on a concrete walkway. The day of the incident was Captain Wilmot's 47th birthday.

After being helped up by the other fire investigator, Captain Wilmot said that he had fallen on his camera. The lens of the camera had created an impression on his torso. Captain Wilmot told the other fire investigator that he thought he had bruised a rib.

Captain Wilmot reported the fall to his employer and saw a doctor on the day of the fall. Pain medication was prescribed and he was released to restricted duty.

When he fell, Captain Wilmot injured his spleen and developed a stress ulcer. The extent of his injuries were not known to Captain Wilmot. The ulcer perforated and released bowel contents into Captain Wilmot's abdomen, leading to infection. Captain Wilmot became ill and was eventually admitted to the hospital. He died on August 9, 2000. The cause of death was due to multiple organ system failure due to peritonitis with severe hypotention, ischemic necrosis of the liver and kidneys due to blunt force trauma of the left chest wall with splenic hematomas, and a perforated stress ulcer.

The cause of the original structure fire was listed as suspicious. Four children were later arrested and charged with arson and criminal damage to property.

Additional information about this incident may be found in NIOSH Fire Fighter Fatality Investigation F2000-37.

Firefighter Fatalities in the United States in 2000

August 2, 2000
Richard Stark, Ambulance Captain
Age 62, Volunteer
Thornhurst Volunteer Fire and Rescue Company, Pennsylvania

Captain Stark and members of his fire department responded to provide care for an elderly female in respiratory arrest. The removal of the patient through her house to the ambulance had been very difficult. Captain Stark climbed into the ambulance and sat in the Captain's chair. At this time he experienced a heart attack. The ambulance was already enroute to a rendezvous point to meet paramedics. The ambulance continued to the rendezvous point, paramedics there treated Captain Stark. Despite the efforts of members of Captain Stark's company and others, Captain Stark died.

August 3, 2000
Phillip Arthur "Pip" Conner, Seasonal Firefighter
Age 29, Wildland Seasonal
National Park Service, Lake Meade National Recreation Area, Nevada

Firefighter Conner was a passenger in a Bell Ranger helicopter that was preparing to return to base for the night after helping to fight the Charlie fire. As the helicopter lifted off, it veered violently to the right and the rotor blades made contact with the ground. The helicopter came to rest back on its skids and the pilot shut the engine down.

Firefighter Conner was severely injured and died that same day as the result of trauma. The other passenger and a crew member on the ground that came to their aid were injured Firefighter Conner was wearing a seat belt at the time of the accident.

Additional information about this aircraft accident may be obtained at the National Transportation Safety Board web site under report number LAX00GA286.

Firefighter Fatalities in the United States in 2000

August 6, 2000
Bradley Scott Pierce, Firefighter/Paramedic
Age 27, Career
Saint Charles City Fire Department, Missouri

Firefighter/Paramedic Pierce had finished a 24-hour shift and was participating in fire department approved physical fitness activities in the basement of his fire station. Firefighter/Paramedic Pierce was alone in the workout area. During the shift, Firefighter/Paramedic Pierce had responded to an emergency medical call and a false alarm. At some point during his workout, Firefighter/Paramedic Pierce suffered a heart attack. Other firefighters discovered him in seizures and provided immediate medical help, to no avail. Firefighter/Paramedic Pierce was a member of his department's Combat Challenge Team.

August 9, 2000
Lisa Ann Farrow, Firefighter
Age 30, Volunteer
Engelhard Fire and Rescue, North Carolina

Firefighter Farrow had provided support at the scene of a fire that was confined to food on the stove. As she was returning equipment to the apparatus at the conclusion of the incident, she collapsed.

EMS assistance was on-scene and provided aid immediately. Firefighter Farrow was transported to the nearest hospital, a 50-mile trip. Shortly after her departure from the scene, Firefighter Farrow went into cardiac arrest. The cause of death was listed as acute hypoxia due to pulmonary edema.

Firefighter Farrow had complained of the heat that day. The temperature was over 90° with significant humidity. Firefighter Farrow had a history of heart problems.

Firefighter Fatalities in the United States in 2000

August 11, 2000
James Alan Burnett, District Forester
Age 51, Wildland Career
Department of Agriculture, Forestry Services, Oklahoma

Four times during the summer of 2000, District Forester Burnett received leave from his full-time job in Oklahoma to work as a temporary firefighter for the Federal government. He had served two assignments in Florida and one assignment in Louisiana. On August 2, 2000, he accepted another assignment as the engine boss of an Oklahoma contract engine and was eventually assigned to the "Kate's Basin" fire in Wyoming. District Forester Burnett's engine was assigned to assist local firefighters with a burnout operation. As District Forester Burnett was sizing up the fire line, a sudden wind caused the fire to "blow up." District Forester Burnett attempted to start the pump on his engine to protect his position but was unable to start the pump. District Forester Burnett attempted to reach a safety zone and attempted to deploy his fire shelter, but was unsuccessful. District Forester Burnett was wearing brush gear at the time of his injury. District Forester Burnett died of burns. He was the first Oklahoma Department of Agriculture, Forestry Services, employee to lose his life fighting a fire.

August 12, 2000
Logan D. Fields, Assistant Chief
Age 51, Career
Hazard Fire Department, Kentucky

Assistant Chief Fields was on-duty in the fire station. He was walking from the bunkroom to the hallway when he fell to the floor, the victim of a heart attack. He died later that day.

August 13, 2000
Lester Lee Shadrick, IFR Captain
Age 53, Wildland Contract
ERA Aviation, Lake Charles, Louisiana   under contract to the Bureau of Land Management, working in Cold Springs, Nevada

IFR Captain Shadrick was working the Twin Peaks fire northeast of Fallon, Nevada. He was piloting a Bell 412 helicopter and was the lead chopper in a flight of two helicopters preparing to make a water drop on a fire-involved ridgeline. The helicopter was carrying a bambi bucket suspended below the aircraft.

As IFR Captain Shadrick approached the ridgeline, his aircraft made a sudden 90-degree left (descending) turn and impacted the mountainous terrain. No radio communication was received from his helicopter after the turn and before the crash. IFR Captain Shadrick was killed instantly.

Firefighter Fatalities in the United States in 2000

August 13, 2000
Warren (J.C.) Smith, Private
Age 28, Career
Indianapolis Fire Department, Indiana

Private Smith was participating in dive training exercises at a local quarry. Private Smith and his company were simulating the rescue of a drowning child in 70 feet of water. Private Smith had been a certified team diver for 2 years. Private Smith failed to surface with his buddy and rescue attempts were commenced immediately. CPR was started in the water and continued during a boat ride to shore. Paramedics from a local rescue squad provided aid, and Private Smith was transported by helicopter to a local hospital where he was pronounced dead. Private Smith died of barotrauma, a condition caused when a diver rises to the surface of the water too quickly and suffers internal injuries as a result of gas expansion during the ascent.

Additional information about this incident may be found in NIOSH Fire Fighter Fatality Investigation F2000-38.

August 13, 2000
Grant F. Trick, First Assistant Chief
Age 49, Volunteer
Canton Fire Department, Pennsylvania

First Assistant Chief Trick was in his fire station gathering firefighters to perform a controlled burn of some brush near a residence. The burn had been planned in advance and had been requested by a local resident. As he prepared for the activity, First Assistant Chief Trick suffered a heart attack. Other firefighters in the station summoned paramedics, and First Assistant Chief Trick was transported to a hospital. He died later that day.

Firefighter Fatalities in the United States in 2000

August 14, 2000
James Robert Renfroe, Assistant Chief
Age 47, Volunteer
Dallas County Fire & Rescue Services, Texas

Assistant Chief Renfroe responded in a mini-pumper to a fire that involved a 240-foot long wooden railroad trestle. He worked on the scene for approximately 6 hours acting as a sector officer and pump operator.

Near the conclusion of the incident, the incident commander instructed him to bring the mini-pumper back to the station to provide coverage for his area. Assistant Chief Renfroe got into the vehicle, started it, and then turned to the passenger and told him that he was not felling well. EMS personnel on-scene were called and Assistant Chief Renfroe collapsed. Firefighters found no pulse or respiration and CPR was begun. Assistant Chief Renfroe was transported to a local hospital by air ambulance.

Despite efforts at the scene, in the air ambulance, and at the hospital, Assistant Chief Renfroe was pronounced dead at the hospital. The cause of death was listed as atherosclerotic cardiovascular disease. Assistant Chief Renfroe was an Equipment Supervisor for the City of Dallas Fire Department.

The railroad trestle fire was believed to be accidental, caused by a passing train.

Firefighter Fatalities in the United States in 2000

August 17, 2000
Robert Wayne Crump, Firefighter
Age 37, Career
Denver Fire Department, Colorado

Firefighter Crump and members of his squirt company were directing traffic away from an area that had been flooded by a very heavy rain. Firefighter Crump was wearing full structural protective clothing including a protective coat, protective trousers, and a helmet. According to the police report, 2½ inches of rain had fallen in the 2 hours prior to this incident.

As the firefighters were working, a woman who was attempting to cross a flooded area stalled her car in the high water and was attempting to walk to a nearby bank to make a phone call. She attempted to cross a rain-filled ditch and fell into the water. She became stuck in a pool of water that covered a culvert but was able to grab onto a pipe to prevent being drawn underwater. Unbeknownst to anyone on the scene, the ditch led to a 64-inch concrete drainpipe that was not equipped with any type of grating.

Firefighter Crump and another firefighter were summoned by the calls of citizens who saw the woman's predicament. Both firefighters entered the water to rescue the woman. As they made their way to the woman, Firefighter Crump was immediately drawn under the water. Citizens assisted the other firefighter from the water, he returned to rescue the woman, and then turned his efforts toward attempting to locate Firefighter Crump.

Approximately 5 hours later, Firefighter Crump's body was located by a police officer near an outlet of the stormwater system. His cause of death was listed as drowning.

Firefighter Fatalities in the United States in 2000

August 20, 2000
John Paul "J.P." Pritchett, Sr., Forestry Crew Chief
Age 56, Career – Wildland
Mississippi Forestry Commission, Webster County, Mississippi

Forestry Crew Chief Pritchett and a forester from the wood products company that owned the land that was on fire were teamed together. The assignment for the pair was to use a tractor-plow (operated by Forestry Crew Chief Pritchett) to cut a fire break to tie in the rear and contain the right flank of the fire. As the tractor-plow worked, the brush between the tractor line and the burned area was set ablaze by the forester using a drip torch so that future spread could be prevented.

The backfire became too intense, so the decision was made to stop the backfire part of the operation. As the forester continued to follow the tractor-plow, he encountered a bee's nest that had been plowed through by the tractor-plow. The forester, who was allergic to bee stings, made attempts to get through the area but was forced to return to the road to avoid them.

About the same time, Forestry Crew Chief Pritchett made a turn toward the fire in an attempt to locate the perimeter. Visibility was poor due to intense undergrowth and smoke. He inadvertently positioned himself in front of a finger of the fire that was making a rapid run. By the time he saw the crowning head fire rolling toward him, it was too late for a retreat. He used his dozer to create a safety zone. He laid face down in the center and covered himself with dirt in an attempt to protect himself as the fire passed over him. Forestry Crew Chief Pritchett was exposed for about 15 seconds. Forestry Crew Chief Pritchett rose from the ground, extinguished a small fire involving his tractor-plow, and drove the tractor-plow out to a point where he met some other firefighters. He sustained second and third degree burns to his arms, back, neck, and face.

Forestry Crew Chief Pritchett was transported to a local hospital by a local police chief and later transferred to a burn center where he was treated for his injuries. He died suddenly and unexpectedly on September 3, 2000, two weeks after his injury. The cause of death was listed as massive bilateral bronchial pneumonia as the result of thermal burns with hospital immobilization.

Forestry Crew Chief Pritchett was either not equipped with or failed to use a fire shelter. The County Chief Medical Examiner's statement strongly recommended that all Mississippi Forestry Commission wildland crews be equipped with appropriate fire retardant/resistant protective clothing. The medical examiner stated that Forestry Crew Chief Pritchett would likely not have sustained his specific burn injuries had he been wearing protective equipment.

The County Line fire eventually consumed 288 acres. This fire and several others in the area were caused by arson.

Firefighter Fatalities in the United States in 2000

September 4, 2000
Earnest Otis Ousley, Lieutenant
Age 49, Career
Roselle Fire Department, New Jersey

Lieutenant Ousley was assigned for the day to dispatch duties. While dispatching an alarm, Lieutenant Ousley suffered severe shortness of breath and chest pains. Firefighters treated Lieutenant Ousley, paramedics were called, and Lieutenant Ousley was transported to the hospital. Approximately 6 hours after he became ill, Lieutenant Ousley died. The cause of death was listed as heart failure.

Lieutenant Ousley had been hospitalized with shortness of breath a month before his death but had been released to limited duty.

September 4, 2000
Daniel H. Yaklin, Lance Corporal
Age 21, Career Military
Marine Corps Air Station Yuma, Arizona

Lance Corporal Yaklin was performing morning checks on his assigned Airport Rescue Fire Fighting (ARFF) vehicle. As a part of the pump test, he was flowing a handline. As he was operating the handline, he was struck by another ARFF vehicle and killed. According to the commanding officer, horseplay and excessive speed were involved in the incident. Lance Corporal Yaklin was a native of Maumee, Ohio.

September 5, 2000
Howard William Van Hoy, Assistant Chief
Age 67, Volunteer
Austin Volunteer Fire Department, North Carolina

Assistant Chief Van Hoy was operating a pump at a live fire training exercise being conducted by his department and one other fire department. Approximately 1½ hours into the exercise, Assistant Chief Van Hoy fell to the ground, the victim of an apparent heart attack.

CPR was started immediately by other firefighters, and Assistant Chief Van Hoy was transported to a local hospital. He was pronounced dead shortly after his arrival. Assistant Chief Van Hoy's son Billy is the fire chief of his department.

Firefighter Fatalities in the United States in 2000

September 7, 2000
Michael Robert Fossett, Crew Chief
Age 45, Wildland Career
North Carolina Division of Forest Resources

David Timothy Newman, Pilot
Age 40, Wildland Career
North Carolina Division of Forest Resources

Crew Chief Fossett and Pilot Newman were travelling to a public education event in their 1974 UH-1H Huey helicopter. As the aircraft entered Balsam Gap, heavy fog was encountered. The pilot radioed that he was going to land the helicopter until the fog lifted. Shortly after the radio transmission, a rotor struck a tree about 20 feet from the top of a mountain. The rotor was destroyed, and the helicopter came to rest in an inverted position and caught fire.

The cause of death for Crew Chief Fossett and Pilot Newman was listed as massive head and body trauma.

Additional information related to this incident may be found in National Transportation Safety Board Accident Investigation MIA00GA264.

Firefighter Fatalities in the United States in 2000

September 11, 2000
Michael Kenneth Yahraus, Firefighter Paramedic
Age 32, Career
Sarasota County Fire Department, Florida

Firefighter Paramedic Yahraus was participating as a SWAT-medic member of a police SWAT team. The team was practicing high-risk traffic stops. Firefighter Paramedic Yahraus was the driver of one of the police vehicles used in the simulation and was standing on the driver's side of his car after the stop had been made. Another officer, who was playing the role of the suspect in this situation, exited his vehicle and then turned and fired a single shot at the officers who were acting in the role of police officers. The weapon used was a 38-caliber pistol loaded with blanks. The firing of the blank dislodged a lead plug that was installed in the barrel of the training weapon. The lead plug broke the windshield, ricocheted off the window post, and struck Firefighter Paramedic Yahraus in the left eye area.

Officers began to provide first aid to Firefighter Paramedic Yahraus while EMS resources were summoned. Upon their arrival, Sarasota County paramedic firefighters provided ALS care and transported Firefighter Paramedic Yahraus to a helicopter landing site for his trip to the hospital.

An investigation revealed that blank cartridges should not have been used in the training weapon. Gas expelled by the blank when it is fired and debris such as wadding that is in the blank can create pressure and force the lead plug out of the gun. The proper cartridge for use in the training weapon would have been a primer round.

Despite all efforts, Firefighter Paramedic Yahraus died on September 12, 2000. Firefighter Paramedic Yahraus was in the final week of a 5-month law enforcement training program.

September 17, 2000
Robert Wilson Humphrey, Firefighter
Age 62, Volunteer
Maryland Line Volunteer Fire Company, Maryland

Firefighter Humphrey responded to the scene of a motor vehicle collision in his personal vehicle. He parked his car on the right shoulder of the highway and began to cross the road to assist a battalion chief who had already arrived on the scene. As Firefighter Humphrey crossed, a mid-sized sedan struck him. Firefighters arriving in response to the original incident assisted with the treatment of the original accident victim and Firefighter Humphrey.

Firefighter Humphrey and the victim of the original accident were transported to the hospital by helicopter. Firefighter Humphrey died later that day in surgery. The cause of death was listed as a result of multiple trauma.

Firefighter Fatalities in the United States in 2000

September 19, 2000
George David Butler, Assistant Chief
Age 47, Volunteer
Idalou Volunteer Fire Department, Texas

Assistant Chief Butler and members of his department responded to a truck rollover that required extrication. Assistant Chief Butler operated the department's air bags and was successful in lifting the truck off the driver so that the extrication could be completed. Shortly after the truck driver had departed the scene by ambulance, Assistant Chief Butler collapsed of an apparent heart attack. Other firefighters began CPR immediately, and Assistant Chief Butler was transported to a regional hospital where he was pronounced dead 2 hours later. Assistant Chief Butler had no previous history of major illness.

September 21, 2000
Bernard D. (Pete) Scannell, Fire Police Captain
Age 70, Volunteer
Waterloo Fire Department, New York

Fire Police Captain Scannell was driving a rescue truck to the scene of a reported car fire. As the unit responded, Fire Police Captain Scannell was struck with a heart attack. The rescue truck left the roadway, jumped a curb, and came to a stop in a small flower bed. Other firefighters immediately removed Fire Police Captain Scannell from the truck and began CPR. An ambulance arrived shortly thereafter and applied a defibrillator. After a shock was administered, a pulse was detected.

Fire Police Captain Scannell was transported to the hospital where he was pronounced dead about an hour later.

Firefighter Fatalities in the United States in 2000

September 24, 2000
Kevin Scott Harshbarger, Firefighter/Secretary
Age 36, Volunteer
Scenic Loop Volunteer Fire Department, Texas

Firefighter/Secretary Harshbarger and members of his department responded to a structure fire in a residence. The fire was in the attic area. Firefighters made an attempt at an interior attack but were forced from the building by extreme heat and smoke. The order was given to open the roof for ventilation.

Firefighter/Secretary Harshbarger and another firefighter went to the roof of the structure to cut a hole. As the hole was being cut, Firefighter/Secretary Harshbarger fell through the roof into the main body of fire. Firefighter/Secretary Harshbarger was not wearing an SCBA. The cause of his death was listed as smoke and soot inhalation.

September 27, 2000
Paul Antonio Lyndell Husband, Sr., Firefighter
Age 33, Career
Mobile Fire-Rescue Department, Alabama

Firefighter Husband and members of his ladder company were dispatched to provide vehicle extrication services at a motor vehicle accident with injuries. As the ladder apparatus was leaving the station to respond to the emergency, Firefighter Husband attempted to board the apparatus. Firefighter Husband chased the apparatus on foot as it crossed two lanes of traffic and made a left turn. He had a hold of a handle near the cab on the driver's side when he lost his grip, fell, and was run over by the apparatus.

The members of the ladder company provided emergency medical care and additional assistance was called. Firefighter Husband was transported to the hospital by ambulance. He was pronounced dead about a half-hour after the incident. Firefighter Husband was working an overtime shift. The cause of death was listed as multiple blunt force injuries.

Additional information about this incident may be found in NIOSH Fire Fighter Fatality Investigation F2000-41.

Firefighter Fatalities in the United States in 2000

October 1, 2000
Thomas G. Gotkowski, Captain
Age 55, Volunteer
Tinley Park Volunteer Fire Department, Illinois

Captain Gotkowski and his engine company were assisting the police department with the ventilation of a condominium. A resident of the home had passed away of natural causes two weeks prior and had not been discovered until that day, ventilation of the condominium was needed. Captain Gotkowski assisted with the placement of a ventilation fan and assisted with the repositioning and stacking of fans. Captain Gotkowski felt ill and was sitting on the back of an engine. He walked to a nearby ambulance where it was determined that he was experiencing a heart attack. He was transported to the hospital where he suffered a fatal heart attack.

This incident was the fifth call of the shift for Captain Gotkowski. According to the fire chief, Captain Gotkowski was the first line-of-duty death for the department since it was founded in 1901.

October 8, 2000
Albert Roger "Bo" Rathbun, Firefighter
Age 69, Volunteer
Sundance Volunteer Fire Department, Wyoming

Firefighter Rathbun was severely burned while cutting a fire break with hand tools. A change of wind "blew up" a pile of smoldering debris at a wildland fire in an area that was thought to be safe. Firefighter Rathbun attempted to outrun the advancing flames, but Firefighter Rathbun fell and was severely burned. Firefighter Rathbun was transported to a burn unit at a hospital in Greeley, Colorado with third degree burns over 40 percent of his body and second degree burns over 10 percent. Firefighter Rathbun suffered a stroke while in the hospital. Despite treatment for his injuries, Firefighter Rathbun died on November 8, 2000.

In an interview that he gave to a local newspaper the day before the fire, Firefighter Rathbun said that he had fought his last fire and that he was retiring from his department. The day after the interview, Firefighter Rathbun and his son were working on their ranch when they saw smoke, responded, and helped to control the fire.

Firefighter Fatalities in the United States in 2000

October 10, 2000
Richard J. LeClair, Captain
Age 53, Career
Federal Fire Department San Diego, California

Captain LeClair had returned to duty after a 6-month battle with cancer. During his fourth work day since his return, Captain LeClair became ill with flulike symptoms on-duty and was transported to a hospital by helicopter. Captain LeClair died the next day as a result of internal hemorrhaging.

October 13, 2000
David C. Fitzgerald, Firefighter
Age 63, Career
Somerville Fire Department, Massachusetts

Firefighter Fitzgerald and his ladder company responded to assist with treatment and cleanup at a collision involving a tractor trailer and a recycling truck. Firefighter Fitzgerald assisted with patient treatment and packaging and then assisted other firefighters as 55 bags of absorbent were distributed over a fuel spill. The incident lasted for over 2 hours. At the scene of the collision, Firefighter Fitzgerald complained of shoulder pain, but dismissed it as a strain.

He collapsed at the fire station after another emergency incident, due to a heart attack. Other firefighters attempted to revive him. Firefighter Fitzgerald was rushed to the hospital by ambulance but was pronounced dead on arrival.

October 16, 2000
Kenneth T. Miller, Captain
Age 65, Volunteer
Cape Charles Volunteer Fire Company, Virginia

Captain Miller was the backup person on a 2-½ inch line that was being operated on a well-involved three-story wood frame residence. Captain Miller collapsed and medical care was immediately initiated by the firefighter that had been on the nozzle. Captain Miller was treated by EMS personnel on the scene and transported to the hospital. He was later pronounced dead at the hospital, the victim of a heart attack. The fire was caused by arson.

Firefighter Fatalities in the United States in 2000

October 26, 2000
James Reavis, Captain
Age 69, Volunteer
North Stone Northeast Barry County Fire Protection District, Missouri

Captain Reavis was the first firefighter to arrive at the scene of a residential fire. As fire apparatus began to arrive, Captain Reavis assisted with stretching lines and setting up equipment. Once things were underway, Captain Reavis drove his personal vehicle to the fire station to retrieve the department's tanker (tender). The station was a short distance from the fire. Captain Reavis was stricken with a heart attack and ran into a parked pickup truck that belonged to a firefighter working on the scene.

Several firefighters were diverted from the fire to provide aid to Captain Reavis. Despite their efforts, Captain Reavis was pronounced dead at the scene.

October 31, 2000
Robert M. Samanas, Firefighter/Paramedic
Age 52, Part-Time
Rural/Metro Fire Department, Bethlehem Steel, Chesterton, Indiana

Firefighter/Paramedic Samanas had completed his yearly physical agility test and stopped to take a break. About 40 minutes after completing the test, Firefighter/Paramedic Samanas returned to assist other firefighters with the test. At this point, he became short of breath. He was placed on oxygen, started feeling better, and then began to experience chest pain. ALS cardiac procedures were started; however, Firefighter/Paramedic Samanas collapsed before a monitor defibrillator could be attached. Firefighter/Paramedic Samanas was transported to a local hospital where he later died. No autopsy was performed.

Firefighter Fatalities in the United States in 2000

November 2, 2000
Jared Conner McCormick, Firefighter
Age 19, Volunteer
Bono Fire Protection District, Arkansas

Firefighter McCormick attended the weekly meeting and work night at his fire department. It was determined that a piece of apparatus was in need of fuel. Two firefighters, including Firefighter McCormick as the passenger, mounted the truck and headed for a fuel station. On the way, the apparatus stalled and could not be restarted. Firefighter McCormick radioed the fire station, told them about their vehicle trouble, and requested assistance. Other firefighters brought another fire truck and the chief's pickup to the location of the broken truck. After the truck was removed from the roadway, an attempt was made to boost or jump start the fire truck using the chief's vehicle. When this failed, it was decided that the broken truck would be towed back to the station by the other fire truck. The chains needed for the job were aboard the other fire truck. At this time, the broken truck and the chief's vehicle were off the road and the fire truck that was to tow the broken truck back to the station was parked across the street due to construction in the area. The area was dark, and the four-way flashers on the chief's vehicle and the working fire truck were in operation.

Firefighter McCormick began to cross the road to retrieve the chains. Firefighter McCormick signaled to an approaching minivan to stop. As his attention was focused on the minivan, a tractor trailer that approached from the other direction struck him. Firefighter McCormick was thrown into the path of the minivan and was struck a second time.

Firefighters on the scene rushed to Firefighter McCormick's aid and an ambulance was requested. An ambulance arrived on the scene within 7 minutes, and Firefighter McCormick was transported to the hospital. He was pronounced dead at the hospital. The cause of death was listed as massive blunt trauma. Firefighter McCormick's father is a career lieutenant in nearby Jonesboro.

Firefighter Fatalities in the United States in 2000

November 2, 2000
Gail Lynne VanAuken, Firefighter
Age 41, Volunteer
Overisel Township Fire Department, Michigan

Firefighter VanAuken was the passenger in a tanker (tender) responding to a mutual aid structure fire involving a turkey farm. Firefighter VanAuken's husband was driving the 2,000-gallon tanker with lights and siren in operation. As the apparatus approached an intersection, a pickup truck approaching the intersection from the other side street appeared to be yielding the right of way to the tanker. The tanker slowed before going through the stop sign. As the tanker proceeded through the intersection, it was struck by the pickup at the left rear axle. The tanker rolled over, the water tank separated from the chassis, and both firefighters were trapped in the cab.

Firefighters from other departments responding to the fire came upon the accident scene and provided aid. Both firefighters were extricated from the cab and transported to the hospital by ambulance. The extrication took about 30 minutes. The injuries to the other firefighter and the driver of the pickup were not life threatening. Firefighter VanAuken received crushing blunt force chest injuries; her cause of death was listed as mechanical and positional asphyxiation. Firefighter VanAuken was the first firefighter fatality in the history of her department.

November 9, 2000
James G. Hill, Sr., Firefighter/Safety Officer
Age 67, Volunteer
Daingerfield Volunteer Fire Department, Texas

Firefighter/Safety Officer Hill responded with other members of his department to a mutual aid structure fire involving a mobile home. Firefighter/Safety Officer Hill assisted with support duties on the fireground and moved inside of the mobile home to assist other firefighters that were performing overhaul. There was a very light smoke condition inside the mobile home. Firefighter/Safety Officer Hill became short of breath and on-scene paramedics began treatment. While enroute to the hospital by ambulance, Firefighter/Safety Officer Hill suffered a heart attack. He was revived but suffered another heart attack in the hospital and died.

Firefighter Fatalities in the United States in 2000

November 15, 2000
Kenneth W. Kerr, Firefighter
Age 44, Career
Fire Department City of New York, New York

Firefighter Kerr and members of his engine company had just returned from fighting a stubborn fire in an elevator cab in a six-story building. At the scene, Firefighter Kerr told other firefighters that he did not feel well but refused medical aid. When his company returned to quarters, Firefighter Kerr spent some time with other firefighters in the kitchen and then headed for the shower. He was found collapsed in the shower by other firefighters. Medical treatment was administered immediately by other firefighters, but Firefighter Kerr died of a heart attack.

November 16, 2000
Kyle Allen Hendrick, Firefighter
Age 19, Volunteer
Gott Volunteer Fire Department, Kentucky

Firefighter Hendrick was the driver of a 1,500-gallon fire department tanker (tender) participating in a water shuttle drill. The passenger in the truck was a 17-year-old trainee. Neither Firefighter Hendrick nor the trainee was wearing a seat belt.

As the tanker traveled down the road, the vehicle's right wheels dropped off the roadway. Firefighter Hendrick overcorrected to the left and came back on the road, riding the centerline. He corrected again and went off the roadway on the right-hand side. The tank separated from the vehicle, and the cab came to rest on its top. Firefighter Hendrick was partially ejected from the vehicle. The trainee was fully ejected from the vehicle.

Firefighter Hendrick was removed from the vehicle and transported to the hospital by ambulance. He was pronounced dead at the hospital approximately 1 hour after the collision. The trainee was severely injured.

November 16, 2000
Phillip Dewey Smith, Driver/Operator Engineer
Age 49, Career
Department of Defense Fire Department, Fort McPherson and Fort Gillem Fire and Emergency Services, Georgia

Driver/Operator Engineer Smith was a participant in wildland fire fighting training. He was the crew boss and helped his crew dig a firebreak as a part of the exercise. As the work was completed, Driver/Operator Engineer Smith fell to the ground and went into seizures. Firefighters that had been involved in the exercise provided immediate care and an ambulance was summoned. Driver/Operator Engineer Smith was transported to the hospital where he died of a heart attack.

Firefighter Fatalities in the United States in 2000

November 17, 2000
Thomas J. Hazaz, Fire Police Lieutenant
Age 69, Volunteer
Tunkhannock Township Volunteer Fire Company, Pennsylvania

Fire Police Lieutenant Hazaz responded to the scene of a motor vehicle accident with members of his department. When he arrived on the scene in his personal vehicle, Fire Police Lieutenant Hazaz received orders from the fire chief by radio. As he passed the scene enroute to his assignment, he waived the fire chief over to his pickup. The chief opened the door of the pickup and repeated his orders. Fire Police Lieutenant Hazaz waved to acknowledge the order and placed his hands on the wheel. The chief closed the pickup's door and noted that the vehicle did not move. The chief opened the door and discovered that Fire Police Lieutenant Hazaz was suffering a heart attack.

Firefighters removed Fire Police Lieutenant Hazaz from his pickup, CPR was started, and an ambulance was called. The ambulance that was on scene for the initial accident had departed for the hospital. Despite efforts on the scene and on the way to the hospital, Fire Police Lieutenant Hazaz was pronounced dead at the hospital. The cause of death was listed as athersclerotic cardiovascular disease. Fire Police Lieutenant Hazaz's death came on the eighth anniversary of his appointment to the fire department.

November 25, 2000
Marvin Maurice Bartholemew, Professional Firefighter II
Age 30, Career
Pensacola Fire Department, Florida

Firefighter Bartholemew responded as a member of an engine company to a report of a residential fire. Upon arrival on the scene, the first company officer reported a working fire with approximately 50 percent of the building involved. Firefighter Bartholemew was assigned to join the crew of a rescue and perform a search of the structure. A handline was stretched by the search crew and carried into the structure. Five to ten minutes after arrival, the company officer from the rescue realized that fire was spreading behind them. He ordered his crew to abandon their efforts and leave the house. All three firefighters headed for the exit as the flashover occurred.

The company officer and the firefighter from the rescue emerged from the structure, both were burned. Firefighter Bartholemew was not with them. The company officer reported Firefighter Bartholemew missing. At least four searches were completed before Firefighter Bartholemew was located. His body was located approximately an hour after the flashover. He had apparently become disoriented and ended up in the kitchen at the back of the house. The cause of death was listed as asphyxia due to smoke inhalation. The carboxyhemoglobin level in Firefighter Bartholemew's blood was 69.5 percent.

The fire was caused when a pan caught fire on top of the kitchen stove and extended. The occupants of the house had evacuated prior to the arrival of the fire department.

Firefighter Fatalities in the United States in 2000

November 26, 2000
Daniel I. King, Firefighter
Age 21, Volunteer
Cliffside Park Fire Department, New Jersey

Firefighter King was responding to an automatic fire alarm in his personal vehicle. He was not displaying emergency or courtesy lights, but he was flashing his headlights and honking his horn. As he responded, a vehicle emerged from a side street on his right. Firefighter King swerved into the oncoming lane to avoid the collision, his vehicle began to fishtail, and he hit a transit bus head-on.

Firefighters responded to the scene and extricated Firefighter King from his vehicle. Firefighter King was wearing a seat belt but the force of the crash was too great. He died later that day. The cause of death was listed as internal trauma.

November 29, 2000
Elwood Queen, Firefighter
Age 67, Volunteer
Irvona Volunteer Fire Company, Pennsylvania

Firefighter Queen was the driver of a fire department ambulance that was transporting a cardiac arrest patient to the hospital. As the ambulance was enroute to the hospital, Firefighter Queen experienced a heart attack. The ambulance left the road, hit a utility pole, rolled two and one half times, and ended up on its roof.

A paramedic and two EMTs riding in the ambulance received minor injuries and provided treatment for Firefighter Queen. They were able to revive him on the scene, but Firefighter Queen died the next day. The patient that was being transported expired on the scene.

The ambulance involved in the accident was two months old and it was destroyed. A fire in December of 2000 heavily damaged the Irvona Volunteer Fire Company station and some of its equipment.

Firefighter Fatalities in the United States in 2000

December 1, 2000
George H. Cardozo, Firefighter
Age 80, Volunteer
Westport Volunteer Fire Department, Connecticut

Firefighter Cardozo worked on the scene of a residential structure fire in his role as fire department photographer. At the scene of the incident, Firefighter Cardozo complained of indigestion but refused help from EMS personnel at the scene. He returned home at the conclusion of the incident and suffered a heart attack after midnight

A police officer was the first to reach Firefighter Cardozo's home and applied an AED. Firefighters provided CPR and assisted EMS personnel. Firefighter Cardozo was transported to a local hospital where he was pronounced dead.

The fire was caused by arson. December 1st was Firefighter Cardozo's 50th wedding anniversary. He responded to the fire after the celebration.

December 11, 2000
Edward A. Russ, Firefighter
Age 39, Volunteer
Bethel Volunteer Fire Department, Vermont

Firefighter Russ was on his way to work when he stopped to assist the occupant of a car that had spun out of control and hit a guardrail. Firefighter Russ found that the occupant of the vehicle was not injured severely, and he turned his attention to directing traffic around the car to avoid subsequent collisions. After only a few minutes on the scene, Firefighter Russ was struck by a pickup truck that was travelling at a speed estimated at 70 miles per hour. He was killed instantly.

A state highway truck was also on the scene. The driver had just begun preparations to move the truck to the other side of the road to protect the site of the original collision.

The driver of the vehicle involved in the original collision was intoxicated and was later charged with driving while intoxicated.

Firefighter Fatalities in the United States in 2000

December 17, 2000
Charles E. H. Lauber, Jr., Commissioner
Age 55, Volunteer
Smithtown Fire Department, New York

Commissioner Lauber and other members of his department had just completed the department's annual Christmas parade. Commissioner Lauber was on top of a fire truck attempting to reset the motor of a garage door opener that had malfunctioned. He fell off the top of the truck and suffered a severe head injury.

Commissioner Lauber was transported to the hospital where he died on December 24, 2000.

December 17, 2000
Keith P. Purcell, Firefighter
Age 47, Volunteer
Southold Fire Department, New York

Firefighter Purcell and members of his department responded to a report of a structural fire. As Firefighter Purcell advanced a hoseline toward a fully-involved detached two-car garage, he collapsed.

Other firefighters came to his aid immediately and CPR was started. ALS care and transport to the hospital were provided by members of the Southold Fire Department. Firefighter Purcell was pronounced dead approximately one hour later at a local hospital. The cause of death was a heart attack.

Firefighter Purcell had been diagnosed with leukemia several years prior to his death. Hours before his death, Firefighter Purcell had played the role of Santa Claus at his department's annual Christmas party.

Firefighter Fatalities in the United States in 2000

December 23, 2000
David A. Anderson, Firefighter
Age 43, Career
Manchester Fire Department, New Hampshire

Firefighter Anderson responded with his engine company to a structure fire involving a three-family residence. Firefighters found a working fire upon arrival with reports of people trapped inside. Firefighter Anderson assisted with fire control and search and rescue functions. Two unconscious boys were located and removed from the fire building by firefighters.

After 20 minutes inside the structure, Firefighter Anderson came outside, sat on the rear step of an engine, stood up, and collapsed. Firefighters provided assistance immediately and Firefighter Anderson was transported by ambulance to the hospital. The cause of death for Firefighter Anderson was listed as a heart attack.

Two boys, ages 14 and 17, were also killed in the fire. The 17-year-old boy had reentered the house in an attempt to save his younger brother. An overloaded electrical extension cord caused the fire.

Firefighter Fatalities in the United States in 2000

December 23, 2000
Scott P. Gillen, Lieutenant
Age 37, Career
Chicago Fire Department, Illinois

Lieutenant Gillen and members of his truck company were dispatched to the site of a motor vehicle collision on an expressway to provide a traffic shield with their apparatus and to assist ambulance personnel. Two state police cars were positioned behind the ladder truck in a further attempt to divert traffic.

As the incident was being concluded, Lieutenant Gillen walked around the truck to make sure that everything was ready to go. As Lieutenant Gillen walked, a passenger car ran over a line of flares in an attempt to slip by traffic. The car then struck a tractor trailer, spun, and pinned Lieutenant Gillen between the car and the ladder truck.

Lieutenant Gillen was treated at the scene and then airlifted to the hospital. His legs were crushed in the collision, and he had lost a substantial amount of blood. He died 10 hours later.

The driver of the car that struck Lieutenant Gillen was under the influence of alcohol and was driving on a suspended driver's license. He was later charged with reckless homicide. There were no injuries in the original collision. Lieutenant Gillen had been promoted to lieutenant just two weeks prior to his death.

The Chicago fire commissioner was quoted as saying "I have a hard time calling this an accident, this was a crime, an absolute crime."

Firefighter Fatalities in the United States in 2000

Additional information about many of the firefighter fatalities presented in this appendix is available from the sources below. Where known, the report number for each incident is listed in the appendix along with the incident description. Many reports are available through the mail and the internet.

NFPA International
1 Batterymarch Park
P.O. Box 9101
Quincy, MA 02269
(617) 770-3000
http://www.nfpa.org

National Institute for Occupational Safety and Health (NIOSH)
Fire Fighter Fatality Investigation and Prevention Program
1095 Willowdale Road
Mail Stop P-180
Morgantown, WV 26505-2888
http://www.cdc.gov/niosh/firehome.html
(800) 35-NIOSH

National Transportation Safety Board
Aviation Accident Database
http://www.ntsb.gov/NTSB/Query.asp

(Use the NTSB Accident Number field to search for a particular incident.)

www.ingramcontent.com/pod-product-compliance
Lightning Source LLC
Chambersburg PA
CBHW081139170526
45165CB00008B/2734